Robert Allen is a Belfast-born journalist working for the Irish national media. His previous work includes a biography of Northern Ireland soccer manager Billy Bingham, reports on pollution in Cork Harbour and on the toxic waste trade, and (with Tara Jones) *Guests of the Nation*, published by Earthscan Publications in 1990. He lives in County Mayo.

WASTE NOT, WANT NOT
The Production and Dumping of Toxic Waste

Robert Allen

EARTHSCAN PUBLICATIONS LTD
LONDON

628.42
ALL

Waste Not, Want Not was researched and written with the assistance of Anne Sammon, Fiona Sinclair, Ralph Ryder, Rosemary Vaughan and David Powell. Sources for the book were mostly secondary. Contemporary material included correspondence, minutes of meetings (public and private) and press cuttings. Original documents, reports and books are referred to. Primary sources are mostly anecdotal but retrospective and therefore based on memory. Quotes taken from newspapers and magazines are in the context of the event as it happened. I am particularly grateful to the journalists all over Britain and Ireland who covered these events as they happened.

First published 1992 by

Earthscan Publications Ltd
3 Endsleigh Street, London WC1H ODD

British Library Cataloguing in Publication Data
Allen, Robert
    Waste not, want not: The production and dumping of toxic
    waste.
    I. Title
    363.17

ISBN 1-85383-095-X

Production by Bob Towell
Typeset by Bookman Ltd, Bristol
Printed and bound in Great Britain by
The Guernsey Press Co. Ltd., Guernsey, Channel Islands

13/9/93D

Contents

Acknowledgements

A book of this kind could not be researched and written without the help of many people. Some of these people gave their assistance on the agreement they would not be named. However I am obliged to acknowledge the assistance of the following: British Rail International and all the staff in the Dublin office, the Derry Development Education Centre and in the Women's Environmental Network in London. Paul Johnston, Alastair Hay, Ann Link, Paul Dowding, Andrew Lovett and Tony Garrell shared their scientific expertise; the anti-toxic campaigners, notably Peter MacKenzie, Jim O'Neill, Rosemary Vaughan, Eamonn Deane, Eddie Kerr, Mick Conlon, Sharon Devine, Eric Gingles, Catherine McGinty, Hugh Brady in Derry; Enda Craig, Bev and Mike Doherty, Dan McGuinness in Inishowen; Paul Fleck and Danny Holmes in Limavady; Neil and Fiona Sinclair and Des Scholes in Inverness; David and Denise Powell in Pontypool; Carol and Ralph Ryder in south Wirral; John Wheeler, Sandra Farquhar; Carol Baker; Liz Jeffress; Erica Woods; Jane McNulty; Ray Jackson; Hilda Palmer; Foster Evans; Cathy Russell; William Halkett; Anya Brennan in Earthwatch, Clare O'Grady Walshe and Bev Thorp in Greenpeace (Ireland) and Paul Miligan. Thanks also to Derek Bateman; to Hilda Palmer in the Manchester Hazards Centre, to Ally McLaws (doyen of Scottish investigative reporters), MEPs Llwellyn Smith and Alec Falconer; Tara Lamont in the British Medical Association; Richard Wilie in Du Pont and in the Department of the Environment press office in Dublin, Jim Ellis and Pat Macken. I am also grateful to Jim Carwood, Tim Ryan, John Mulcahy, Paddy Prendiville and Damien Kiberd.

Finally I must mention Ann, Patricia, Grattan, Tomas, Mark, Doug, Lucy, Tom, David, Helen, Orla, Alife, Ruth and Ciara, my editors Neil, Ewan and Sian at Earthscan (and the regulars and barmen in Mulligans, the Waterloo, and, just at the end, Ryans in Cong).

Robert Allen
September 1991

To Ciara

Introduction

Communities need to get involved. The other important thing that people need to remember – apart from when they get that information – is that they probably know as much or maybe even more than some of the experts drafted in to view certain things. I know I get called into certain things; I get asked to give opinions on certain things, someone asks me about something they've been researching for a long time and they've put a lot of stuff together and quite often they know far more about it than I do, yet someone will defer to me because I've done toxicology or something else. In one way I understand it but in another I am increasingly impressed with what people can put together."

Alastair Hay
Department of Chemical Pathology, University of Leeds

Pollution from toxic waste disposal sites affects communities, wildlife, farmstock and the environment. To what extent this pollution actually endangers the health of humans is unclear. Evidence of links between toxic waste disposal and ill-health in humans is sparse. Communities who live beside toxic waste disposal sites are convinced there is a problem. The waste disposal industry has insisted that their "state-of-the-art" incinerators are safe, that their landfill sites are adequately sealed and that their storage, treatment and recovery methods do not endanger the health of employees or local communities. The communities are not convinced. The major concerns are emissions from incinerators, primarily merchant incinerators – which take hazardous waste – and clinical incinerators, which take multifarious hospital waste. Municipal incinerators are not perceived as a major problem despite evidence that they also pose a risk to communities.

In Britain and Ireland there are four private sector high temperature toxic waste incinerators and proposals, at various planning stages, for at least a dozen more. Hospital and municipal incinerators proliferate throughout our cities. Following the removal of Crown immunity from clinical incinerators, waste disposal companies are falling over themselves to control this lucrative market. Public fears that emissions from the incineration process are detrimental to human health have been heightened by the Hanrahan case in Ireland and the controversy over ReChem's operations in Scotland and Wales. The problem for communities and environmentalists has been proof.

Dioxins and furans are commonly mentioned as products of incineration. Although there are 75 types of polychlorinated dibenzo-para-dioxins (PCDDs), the most studied and the most toxic is 2,3,7,8-tetrachlorodibenzo-para-dioxin (TCDD). There are 135 different types of polychlorinated dibenzofurans (PCDFs). The major known sources of dioxins and furans, particularly TCDD, are chemical manufacture and incineration. What is not commonly known and not agreed by scientists, epidemiologists and geographers is the effect these chemical compounds have on human health. Incineration does not destroy the various wastes, it simply creates new chemical compounds; in the language of toxicologists these are known as PICs (products of incomplete combustion). The toxic qualities of the majority of these compounds are unknown.

For years many scientists were content to believe that although dioxins are lethal to certain animals there wasn't enough evidence to show any serious long-term effects on human beings. But this belief is slowly changing. "Our report is about the sources of organochlorines, especially the dioxin-like compounds, in Britain and their likely effects on health, especially pre-birth effects," said the Women's Environmental Network scientist Ann Link about their report on dioxin-like compounds and ill-health, especially in children, pre-and post-birth. "I have been going through a lot of published scientific stuff on toxicity and evidence of effects and I feel that the so-called 'anecdotal' evidence is screaming through all the distanced scientific papers!"

Dioxin emissions may be dispersed by air, wind and water, and they may then be ingested by grazing livestock, lodging in the fatty tissue and milk – thus entering the food chain. Human intake of dioxin-like compounds is largely through food, some scientists estimate as high as 98 per cent of the total intake. Dioxins and furans may also be taken into the human body through the skin and by breathing. In a report for Glasgow Department of Public Health on the effects of a proposed incinerator Dr Graham Watt, lecturer in environmental health, said: "If no information is available about the performance of the incinerator to be installed it would be appropriate to label it a prototype and to recognize its initial operations as an experiment."

This proposal is not misplaced. Chemicals, such as drugs which we take voluntarily, are tested on animals, rigorously tested on humans using double-blind control (in clinical drug trials) and finally given to patients. When we buy or are given a pharmaceutical we can find out what its composition is, its rate of absorption, its effects and side effects, including carcinogenic properties (although all the side effects are not always known. The development of a new chemical compound may take years of research before it can be considered for human use. Exposing communities to toxic emissions could therefore be compared to conducting a drug trial using several unidentified chemical compounds, and without the benefit of the controls used in such trials. To evaluate the epidemiological effects of chemicals we must know what toxins are present and how exposure occurs.

This has bothered epidemiologists, many of whom are concerned about the lack of information. "Most of the epidemiological investigations that have been performed to date have been small, cross-sectional studies in which the exposure of the study population to toxic chemicals has not been adequately characterized," Upton, Kneip and Tonilo of the Institute of Environmental Medicine, New York, stated in 1989. They were concerned that of the thousands of known waste disposal sites in the US, few had "been adequately evaluated with respect to the chemicals they contain, the extent to which they are releasing such agents to the surrounding environment, the degree to which they are thereby exposing human populations to toxic substances, and the extent to which they may have affected the health of those exposed".

In Britain and Ireland the situation is much worse. Very little epidemiological work has been done in this area. According to Tony Gatrell, who co-authored *Burning Questions: Incineration of wastes and implications for human health* with Dr Andrew Lovett in 1989, "inadequate data and lack of information about exposure" is a major problem. Scientists and toxicologists in Britain agree that, compared with other countries, information on the health effects of dioxins and furans is sparse. G.H. Eduljee, a consultant to ReChem – who operate high temperature incinerators at Pontypool, South Wales and at Fawley, in southern England – wrote in 1988 that the dioxin research programme in Britain would be seen "as an inadequate response to public anxiety", but by some "as a careful and pragmatic approach, learning from the experiences of other nations and making better use of its research budget". However, there is no dispute that information about the human health effects of chemical compounds produced by incineration is sparse. What is remarkable about this is that the authorities in Britain have had several excellent opportunities to study the effects of these emissions on humans.

Evidence of the health effects of dioxin is limited to the results of industrial accidents (such as the one at Seveso in Italy in 1976), the effects of "Agent Orange" in Vietnam, exposure to herbicide spraying, laboratory experiments on animals, and more recently a US government-sponsored study of 5,172 chemical workers from 12 factories with occupational exposure to TCDD. Dioxin (particularly TCDD) is known to cause chloracne (a skin disfigurement), headaches, nausea, itching, liver and kidney damage, immunosuppression and neurological damage. Because dioxins are persistent it can take seven years for half of the dioxins in our body to be excreted. Fat loss can release large doses into our system. Dioxins are metabolized in the liver, which releases enzymes to break them down, but the same enzymes damage the liver itself. The thymus gland in babies is particularly vulnerable to dioxin damage, and as babies depend on the thymus for immunity they may develop immunosuppression. Sterility may also be an effect of dioxin poisoning.

Tests on rodents have shown dioxins to be carcinogenic and the US study on the chemical workers revealed TCDD to be a carcinogen. Yet statistics may not be indicative of the true incidence of cancer because of the long latency period between exposure and tumour development. Faulty recording at the time of exposure can also present an unclear picture. (This is what happened at Seveso in 1976 (see Table 10.1, pp.212–214), where under-reporting was widespread.) Industrial accidents such as Monsanto (see Table 10.1) have, since 1949, exposed around two thousand workers to high levels of TCDD. Many of these workers developed chloracne. TCDD is also a by-product of the manufacture of trichlorophenol, which is used in the manufacture of herbicides. The most commonly known of these is 2,4,5-trichlorophenoxyacetic acid (2,4,5-T) which was one of the ingredients of Agent Orange. Although TCDD has been known as a contaminant in the manufacture of trichlorophenol and 2,4,5-T for over 30 years and that exposure to TCDD causes chloracne it has only been in the last three years that studies have conclusively linked dioxin to human illness.

In 1989, while the US study was being compiled, a (West) German epidemiologist claimed that he had established the first clear evidence to link dioxin poisoning to cancer. On 17 November 1953 an explosion at the BASF trichlorophenol factory in Ludwigshafen exposed 122 workers to TCDD. In September 1989 Dr Friedemann Rohleder produced a report at the 9th international conference on dioxins in Toronto, Canada, in which he recorded an unexpectedly high incidence of cancer in the BASF workers. Eight workers, he said, had died from cancer.

This evidence is startling because it has implications for an industrial accident closer to home. In April 1968 an explosion occurred at the Coalite factory in Derbyshire. Since 1965 the fine chemicals unit of the company manufactured 2,4,5-T. Following the accident 79 cases of chloracne were recorded, many severe. Although many of the workers' chloracne cleared up within six months, follow-up studies were seen as inadequate and in a strange incident, the medical records of a doctor who had studied the Coalite workers were stolen from her home. It is not yet known what the long-term effects of the TCDD exposure has been on the Coalite workers (the Health and Safety Executive are expected to report on the health of the exposed workers in 1992), but it is a major concern to people like Ann Link. She has noted that there appears to be a real increase in cancers in older people

> which is not explained by an increase in the number of people surviving to older ages. It may be possible that we are seeing the long-term effects of a reduction in immunity caused by dioxin-like substances present in the body for perhaps forty years. The extreme persistence of dioxin in humans means that, once exposed, we continue to receive low doses from our body fat, thus increasing the chance that some cell-level accident will occur eventually. No animal has been studied for the equivalent of a human lifetime.

This has deeply disturbed communities closest to incinerators. No one has been able to convince them that their long-term health is not under threat. "Unequivocal conclusions about health effects are rare," said Dr Andrew Lovett, who added that the current state of epidemiological and toxicological knowledge was insufficient to placate the fears of communities, though he also said it was his impression that some communities had prejudged the issue. If they have prejudged the issue, and several waste disposal companies believe they have, it is because there is not enough evidence to show that incineration is safe. Communities now live with the fear that they may develop cancer and other diseases if exposed to toxic emissions. In the US there is legislation which caters for community "oncophobia" or fear of cancer. In Britain and Ireland there is just the fear.

That fear has been intensified by the fact that Britain imports large quantities of highly toxic waste, such as polychlorinated biphenyls (PCBs), for high-temperature incineration. In the years between 1981 and 1987 the importation of hazardous waste into England and Wales escalated 15-fold. British waste disposal companies are able to offer high temperature incineration at competitive prices, in other words cheaper than most countries. The value of toxic waste imports to Britain in 1988 was over £700 million. In 1988 and 1989 the profits of the major waste disposal companies doubled, though this upward trend was not maintained in 1990 and 1991. During the seventies and eighties ReChem operated the only high temperature incinerator in Britain that could take solid PCB waste. Despite the fear in communities, the British government does not believe there is a problem, and has said that the importation of hazardous waste is acceptable "provided it is properly regulated at all stages". But this regulation, the communities argue, does not occur. Existing legislation has not aided the regulation process as under it the waste disposal companies report to different regulatory bodies, and the lack of an integrated regulatory system has frustrated local authorities and communities. Yet it is a problem also acknowledged by the waste disposal industry. "Any control system is only as good as its enforcement," the National Association of Waste Disposal Contractors stated in 1990. The communities opposed to toxic waste disposal insist that enforcement has not been carried out because of inadequate resources within the regulatory bodies, and the complexity of the legislation. When Torfaen council attempted to prosecute ReChem it discovered it did not have the legislative power.

The growth of the toxic waste disposal industry has been met equally by the growth of the opposition to it. This opposition has been total because nobody, government, industry, academia, has been able to placate the fear among communities about the consequences of toxic waste disposal. The growth of environmental awareness has helped to focus the issue of toxic waste in the public mind so that dioxin is now an addition to the general vocabulary. The effect of dioxin poisoning is not fully known and until it is there will be community opposition to toxic waste disposal.

Waste Not, Want Not is an attempt to record and explain why this opposition was formed, why incineration is feared by communities, why the waste disposal industry is not trusted by communities, why government and the green movement has failed to address the toxic problem adequately. It is about the effect the production and dumping of toxic or hazardous waste has on ordinary people, on communities who do not necessarily endorse or embrace the so-called NIMBY (not in my back yard) syndrome. It is important to stress that while the general theme of the book is around the issue of toxic waste production, transportation and disposal – by treatment, recovery, landfill or incineration – the book is a chronicle of the events in some of the areas where communities have opposed toxic waste sites. So it is about the community opposition to, and concern over, the production and dumping of toxic waste and the stories are told, as much as possible, from the communities' perspective. Unfortunately this book does not offer solutions – the production of less waste, clean production, new "state-of-the-art" incineration and landfill technology, scientific innovation – simply because that is not its theme. The argument that toxic waste must be dealt with is addressed throughout the book. The communities' argument here is simple: those who produce toxic waste should deal with it themselves. If in Britain and Ireland the solution is incineration, communities do not wish to be used as guinea pigs in this experimental technology. The incineration processes used by the major chemical and waste disposal companies are not proven to be safe, and rhetoric will not convince communities that they are. Incineration is used because it can bring a high return on the initial investment and considerable profit thereafter. It is not used because it is an environmentally sound and healthy method of toxic waste disposal. When government and industry understand this they will understand why there is opposition to toxic waste disposal.

Bibliography

Eduljee, G.H. "Dioxins in the Environment", *Chem. in Brit.* 24 (12), 1223–1226 (1988).

Department of the Environment "Dioxins in the Environment", Pollution paper no 27, HMSO (1989).

Fingerhut, M.A. et al, "Mortality among US workers employed in the production of chemicals contaminated with TCDD", US Department of Health and Human Services, NTIS PB 91–125971 (1991).

Gatrell, A. and Lovett, A., "Burning Questions: Incineration of Wastes and Implications for Human Health", North-West Regional Research Laboroty, Lancaster University, Research Report No 8 (1989).

Hay, A. *The Chemical Scythe* (New York: Plenum Press, 1982); see also *Nature* 284, 2 (1980); 285, 4 (1980) and 290, 729 (1981).

May, G. "Chloracne from the accidental production of TCDD", *Brit. J. Ind. Med* 30, pp. 276–283 (1973).

May, G. "Tetrachlorodibenzodioxin: a survey of subjects ten years after exposure", *Brit. J. Ind. Med.* 39, pp. 128–135 (1982).

Rohleder, F., "Dioxins and Cancer Mortality: Reanalysis of the BASF cohort", Presentation to the 9th International Dioxin Symposium, Toronto, Canada, September 17–22, 1989.

Tschirley, T.H. *Sci. American* 254 (2) pp 21–27 (1986).

Womens Environmental Network, *Chlorine, Pollution and the Parents of Tomorrow*. Contact WEN, Aberdeen Studies, 22 Highbury Grove, London N5 2EA for further details.

PART ONE

COMMUNITIES AGAINST TOXICS

Chapter 1

DERRY AND DU PONT
Still desperately seeking a toxic dump

Enda Craig and Bev Doherty live on the anvil-shaped Inishowen peninsula in Donegal. Locally they are known as "green" people, concerned environmentalists, worried about pollution and its effects on human health and on the natural world. Craig is a member of the Greencastle and Moville Environmental group. Doherty is chair of the Inishowen Environmental group. For years both groups have been "beating their heads off a brick wall" trying to convince their neighbours and anyone who will listen that there is a problem with pollution on the peninsula and in the Foyle lough and river. In January 1991 events shaped by state policy brought their worst fears about local pollution into perspective.[1]

The news that US multinational Du Pont had begun to examine proposals to build a national incinerator at its plant in Maydown, Derry which would take toxic waste from the 32 counties of Ireland shocked Craig and Doherty. It also shocked the people of Derry, who were not known as strong environmentalists. In fact, Craig would have argued, Derry people did not have an opinion on environmental issues. If that was true, and there were many who believed it was, it changed on Monday evening, 21 January 1991 when 120 people crammed into Derry's central library. Enda Craig had spoken alongside three members of Greenpeace and a member of the Cork Environmental Alliance. Earlier in the day the Greenpeace/CEA party had met Derry City Council's Environmental Protection Committee. The following day at a meeting of the council "opposition to any toxic waste incinerator facility was confirmed".[2] If the people of Derry had once been soft on environmental issues that time had gone. By the end of the month, opposition to Du Pont's plans were not only vociferous but total, as groups began to form and voice their own particular objections. As one reporter later put it: "From

Moville to Maydown, from Strabane to Strathfoyle, people living in towns and villages all along the river Foyle, people conscious of the environmental destruction and the risk to health were beginning to ask very awkward questions," as toxicologist Paul Johnston had urged them to do.[3]

A few days before the public meeting in the library, Bev Doherty posed some very awkward questions herself in a letter published in the *Derry Journal*; she also provided some salient facts.

Incinerators do not have a good track record for being pollution free. Indeed, the Dutch government has warned against drinking unprocessed milk from cows grazing near toxic waste incinerators. They have found high levels of dioxins in the milk. The soil around ReChem's Welsh incinerator has been found to have 88 times more than the normal PCB levels. PCBs and dioxins are extremely carcinogenic, and also very persistent once in the environment, and will pollute air, land and sea for miles around if incineration is not maintained at temperatures around 1000 °C. This, in practice, is not often the case.[5]

She told readers of the *Journal* that air and water pollution was already a problem in the area "with claims of fish tainted with chemicals and of various foul odours and health problems, both sides of Lough Foyle when the wind is blowing in the 'wrong' direction". What else, she asked, will blow over to Inishowen or County Derry if the incinerator plans go ahead.[6] The Inishowen Environmental Group and the Greencastle and Moville Environmental Group know about Du Pont and their activities in Derry; they are well informed when they speak of the environment and feel strongly about self-determination, the right to be allowed to make decisions about events which affect their lives. Bev Doherty added: "It has been suggested that ordinary mortals keep out of this debate – let's leave it to the experts to make the right decision. This underestimates our intelligence and the ability of people to research into a subject that effects them deeply. Moreover 'experts' can sometimes have vested interests! Our vested interests are our own health and that of the natural environment.[7]

When it became known that Padraig Flynn, Minister for the Environment in the Republic of Ireland, had met, at the initiation of the company, Du Pont's management at their Maydown complex, news that the US multinational were advanced in their plans for a toxic waste incinerator began to emerge in the Irish national and in the local Belfast and Derry media. Du Pont confirmed that they were "considering an option for treating waste on both sides of the border"; a company spokesman added that he doubted "if the chemical industry in Ireland would need two waste disposal facilities on the one island".[8]

Padraig Flynn was more forthcoming about his visit to Du Pont in an interview with the *Belfast Telegraph* where he explained that the US company had indicated that they were considering the possibility of building "a waste disposal facility" to treat their own waste; they also wanted to take into account, said Flynn, the amount of waste that would be generated in the Republic of Ireland over the coming years. "A certain amount of 'feedstock' would be required to make a waste incinerator project viable" added Flynn, who stressed that there was no agreement between the company and the Irish government. "It may well be that the proposal will have something for us, but we will not know until they proceed with their study a bit further."

Environmentalists and community activists opposed to the siting of a toxic waste incinerator anywhere in the Irish countryside were immediately suspicious of Padraig Flynn's "secret" visit, of Du Pont's motives and of the roles of Jacobs International, the Dublin based US project engineering company, and Minchem, the Dun Laoghaire waste disposal company who have approximately 50 per cent of the Republic's toxic waste disposal market. Flynn had, it appeared, been offered a northern Irish solution to a southern Irish problem. Would he be remembered in history as the environment minister who answered the Irish chemical industry's cries for a solution to the decade-long search for a method of disposing of their toxic waste?

It was in July 1988 that Padraig Flynn set in motion the initiative he believed would solve the problem. Dublin consultants, Bryne O'Cleirigh, were commissioned by the Minister to study the feasibility of a national toxic waste incinerator. Incineration, Flynn said later, in November 1988, at a conference in Cork on waste disposal, was the long-term solution to industry's toxic waste problem in Ireland. But the State, he said, would not rush into a decision to construct an incinerator; the Department of the Environment would need first to wait for the Byrne O'Cleirigh report. When the report was delivered in February 1989, with a recommendation that an incinerator be built and up-and-running by mid-1991, Flynn said he wanted a public debate on the matter, a comment later interpreted by environmentalists as little more than a gesture to those who were really concerned about toxic waste disposal. By September 1989 the Department of the Environment was ready and tenders for the project were invited. Flynn's department aimed for a shortlist of six, got five, and eventually narrowed the tenders down to two consortia; Minchem/Jacobs International and Irish Environmental Services/SARP.[10] The "public" debate had not materialized, neither had a referendum.

By December 1990, when Padraig Flynn visited Du Pont, the Department of the Environment in the Republic of Ireland had not made a decision about who would build the incinerator and where it would be located. The Minister was aware that if Ireland did not have a central facility for the disposal of toxic waste before the trade barriers fell in 1992, with the onset of the EC Single Market, there was another solution.

When he spoke at the Institute of Engineers of Ireland annual general meeting on October 11, 1990 he asked: "How do you control the production and movement of waste – which is both a tradeable commodity and a hazard to the environment – in a Europe without frontiers?" This question, he added, had been successfully addressed by the EC in the form of a three-pronged waste policy; a) the prevention of waste by using clean technologies and processes: b) recycling and c) safe disposal of waste in the nearest and most suitable site relative to its point of production.[11]

Enter the Northern Ireland Department of the Environment. In June 1990, English consultants Aspinwall and Co. Ltd presented their "review of waste disposal in Northern Ireland", which had been commissioned by the NIDoE less than a year earlier. It wasn't until January 1991 that the communities in Derry and Donegal, who were opposed to Du Pont's plans for a "national" toxic waste incinerator, learned that the initiative had in fact come from Aspinwall and Co. who recommended that the NIDoE "undertake joint discussions" with Padraig Flynn's department. These discussions, Aspinwall and Co. indicated, would determine the feasibility of the north being included in the current proposals for high temperature incineration and the widening of proposals to include chemical treatment. "Given our understanding of the situation," Aspinwall and Co. concluded:

> we are extremely sceptical that the market could support a high temperature incinerator in both the Republic and in (the north of Ireland) unless it was the intention of the operators to rely on imported wastes from other countries. In this respect such an arrangement would go along way to adopting the spirit of the EC Waste Strategy which was released in 1989.[12]

Aspinwall and Co.'s assertion that there was not enough room in Ireland for two major incinerators was a claim Padraig Flynn had heard before, from Dublin consultants Byrne O'Cleirigh. The suggestion that there should be a united Ireland for toxic waste was giving Flynn a way out of the problem which has bedevilled successive Irish governments. What nobody in authority in either Dublin or Belfast fully realized is that there was also a united Ireland in opposition to toxic waste disposal.

It is in the context of these two studies that Du Pont's invitation to Padraig Flynn is revealing. While both state authorities in Belfast and Dublin dithered, the private sector, as it had done time and time before, took over the initiative.

As the campaign to oppose Du Pont's plans began to grow in mid-January, the communities in Derry and Donegal realized that they not only had the support of their local authorities and most of their elected representatives but also the support of the local media. The civil servants in Derry City Council had been quick to respond to the public's fears about incineration and entered the debate in a professional manner that surprised some of Derry's more cynical citizens. Du Pont had sent two representatives to address the council's Environment Protection Committee on 15 January, a day after a council "source" had told the *Belfast Telegraph* that the authority was "working in the dark and until we know more no one is prepared to lend even cautious support to this kind of project".[13]

Councillor Gregory Campbell, of the Democaratic Unionist Party (DUP), joined with councillor Hugh Brady, of Sinn Fein, in opposition to Du Pont's plans. Gregory Campbell said his party "would be totally opposed to what would be seen by a considerable number of people as the use of (London) Derry as a dumping ground for wastes from the Republic".[14] Hugh Brady issued a detailed statement from Sinn Fein which articulated both the technical and political arguments against incineration:

Industry favours incineration because once the waste is burned and dissipated into the environment, it is extremely difficult to legally trace resultant health and environmental damage to the original polluter. More importantly, it is extremely difficult to legally prove the origin of pollutants thus scattered over land, water and into the atmosphere. Incineration provides a convenient and liability-free way for industry to transfer responsibility for pollution from the waste management industry to the communities in which incineration is sited. Studies show that exposure (to emissions from incinerators) can cause cancer, birth defects, spontaneous abortion and foetal poisoning. They can also damage the reproductive system, cause sterility, weaken the immune system, cause liver and kidney damage and behavioural problems.[15]

Throughout January Du Pont argued that it had not submitted a planning application for a toxic waste incinerator, that the decision to build an incinerator which would take waste from all over Ireland had not been and would not be made until a feasibility study on the proposal had been completed. Following a report in the *Irish Times* which stated that Du Pont had applied for planning permission to build a toxic waste incinerator at the Maydown complex, the US company revealed that it had applied for permission to build a small lycra waste burner which, said a company spokesman, "would have no negligible impact on the environment [*sic*]". The spokesman added that the company did "not have a project for a hazardous waste burner at this moment in time, and no planning permission has been made at this stage. It is possible there had been some confusion between the two issues".[16] Later that month two community activists visited the offices of the NIDoE in Derry to view the planning documents. They were shown Du Pont's outline planning application for the "installation of a replacement solid waste burner to burn lycra waste" which had been submitted on 5 December 1990. The application, they were told, had been forwarded to the DoE headquarters in Belfast and an Environmental Impact Assesssment (EIS) had been requested from Du Pont. The application did not reveal what the capacity of the burner would be or who would build it. (The activists learned later than Jacobs International would design and build the burner.)

Du Pont had, apart from sporadic appearances in the Belfast and Dublin media, remained quiet about the issues which the communities were beginning to debate in private and public meetings. On Thursday 26 January the company circulated a memo to the Maydown workers. Entitled "Waste Incineration at Maydown – the Facts" the document was subsequently sent to the media and to residents closest to the Du Pont factory. Presented in a question and answer format, the document dealt specifically with the proposed "national" incinerator.

Much of the information in the document was known to the communities; if it said anything it confirmed the fears of many of those opposed to the proposed incinerator – that they were being presented with a *fait accompli*. Du Pont, the document stated, "would not participate in any venture that would be harmful to the neighbouring community or the local environment"; the feasibility study on the proposed incinerator would determine whether the company should proceed. If it did, "full consultations would take place with the appropriate government regulatory authorities, the City council, neighbours and local environmental groups". The Derry media faithfully reproduced the document, and the issue began to dominate the pages of the *Derry Journal* and *Sentinel*, and the feature and news programmes of Radio Foyle.[17]

By the following Friday, 1 February, the word "toxic" featured in the conversations of almost everyone in Derry, particularly in the Derry Development Education Centre (DDEC) which had, because of its role as a community resource, become the headquarters of the campaign against Du Pont's plans. The initial campaign had reached out to embrace Greenpeace and its global struggle to end incineration, but the workers in the DDEC were determined from the beginning that the campaign should be fought by the community and not led by environmentalists from Belfast, London and Dublin. The DDEC's role as a development and education resource enabled it to use its contacts to gather information on incineration and toxic waste disposal.

The communities and the media weren't slow to use the DDEC as a resource. Limavady businessman Bob Parke explained to a Radio Foyle reporter:

> About ten days ago there was a meeting in Derry [library] where we became aware that Du Pont were considering the possibility of building a toxic waste incinerator up in the Maydown facility. There were three of us that came away from that meeting very concerned at this possibility. We're all neighbours living in this area, which is in an immediate downwind zone from the Laydown complex. Frankly we were very concerned and we thought we really ought to share our concerns with our neighbours and other people in the area. We felt the best way to do this could be to call something of a public meeting.

The DDEC had organized the meeting in the library and had provided the speakers for the public meeting called by Parke in the White Horse Inn on Wednesday 30 January. Approximately 150 local residents heard Jim O'Neill of the DDEC tell them that the centre had information on incineration, pollution and health effects of toxic waste disposal. This information, O'Neill said, was available to everyone. Parke told Radio Foyle that the attendance had been more than they expected. "We originally thought we'd be lucky to get 50 people. Quite clearly I think that reflects the very serious concern of a lot of people in this area. They wouldn't be coming out on a cold January night if they didn't have serious and very honest worries about his prospect."[19] The audience were worried, particularly about health effects in Derry since Du Pont had set up in 1960. Several people told about gas clouds floating down the Foyle.

One woman said Du Pont had recently sent out a card explaining what to do in the event of a gas escape. As the anger against Du Pont's past activities and their future plans began to intensify, a local member of the SDLP got to her feet and explained that Du Pont's application for a "lycra waste burner" had been taken out of the hands of the council. She added that, while pressure groups were fine, this sort of thing should be left to the politicians. Shouts and jeers of disapproval clearly indicated that not everyone in the audience agreed with her. Parke reiterated the comment he had made at the start of the meeting: that Du Pont had declined an invitation to address the residents nearest to the Maydown complex because, as he paraphrased it, they had nothing to say to the public since no decision had been made about the incinerator. Parke had earlier read out the document Du Pont had circulated to their workers, the media and to some of their neighbours. Several people, who said they lived in the immediate area, said they had not received the document. The meeting closed with a decision to form a local committee: Campsie Residents against Toxic Emissions (CRATE).

The SDLP believed it could provide the answers the communities in Derry demanded but there were proper channels to go through. SDLP leader, Euro and Foyle MP, John Hume, had that week been asking questions in the House of parliament in Westminster. The Secretary of State for Northern Ireland told Hume in a written answer that in November 1990 correspondence from Padraig Flynn in the Republic's Department of the Environment confirmed that the Irish government "was at an advanced stage in its consideration of a central hazardous waste incinerator to service Irish industry" and that Flynn was aware of Du Pont's proposals.[20]

On the question of public safety and environmental impact the government replied:

The effect of a proposal to site a toxic waste incinerator in Northern Ireland on public safety would be examined by the Department of Economic Development's health and safety inspectorate which enforces the provisions of the Health and Safety at Work (NI) Order 1978. The order requires every employer to ensure, so far as is reasonably practicable, that persons not in his employment who may be affected thereby are not exposed to risks to their health and safety. Regulations prevent the granting of planning permission to proposals of this type unless environmental information has first been considered. Schedule 1 (8) of the Planning (Assessment of Environmental Effects) Regulations (NI) 1989 applies and any planning application for such an incinerator must be accompanied by an environmental statement. These statements are available for public information.[21]

Many anti-toxic campaigners wondered why Hume had remained relatively silent on the issue although it was clear that his party had done their homework and were familiar with the issues, particularly the concern about the company's planning application for a "solid waste burner" to incinerate lycra waste. On this matter Hanley told Hume that this burner was for lycra waste only and that schedule 1 (8) of the Planning (Assessment of Environmental Effects) Regulations (NI) 1989 applied to this application. Hanley confirmed that the company had been asked to submit an EIS as part of the application.[22] On Du Pont's "national" incinerator Hanley told Hume that in October 1990 Du Pont had "advised" the NI Industrial Development Board (IDB), the Department of Economic Development and the Department of Environment about its plans. In December Du Pont confirmed "in writing that an evaluation of the project was under way" and that officials in Belfast would be available for "consultations with the company as appropriate".[23]

Back in Derry the debate was getting hotter and more controversial. Radio Foyle's early morning news programme of Thursday 31 January included a report on the meeting in the White Horse Inn and an interview with Jim Braiden of the Chemical Control Division of the OECD in Paris. To the astonishment of the people in the DDEC, who were beginning to grasp much of the arguments about incineration, Braiden asserted that incineration was safe and that PCBs could be completely destroyed in an incinerator.[24] Jim O'Neill knew this was untrue. He rang Radio Foyle and asked if they would run a counter-claim on the one o'clock news if he could find a speaker. Paul Johnston, the London-based Greenpeace consultant toxicologist, agreed to go on the air and refute Braiden's claims.

On the lunchtime programme Radio Foyle introduced Johnston as a "person technically qualified" to answer questions about incineration and whether it was safe or not. "You cannot guarantee a hundred per cent safety," Johnston said, "an incinerator is an open ended system. It disperses materials to the environment and to the atmosphere. Some of the alternatives which are at a research stage now have the advantage that they can be operated in a closed configuration whereby you have complete control over an emissions to the environment. You don't need to make emissions to the environment. It's the fact that you're using the atmosphere as a dilutant for all these materials that makes incineration so cheap and attractive." Should the prospect of large numbers of new jobs not sway the people of Derry, the radio interviewer asked, if it led to an influx of new industries attracted by a toxic waste incinerator? Johnston replied that the incinerator would create no more than 30 jobs and that it was up to the people of Derry to decide whether they wanted the dirty industries which would be attracted by "the fact that they have an incinerator to dispose of their waste". You don't really want that kind of industry, he concluded.[25]

The Radio Foyle interviewer's question about large numbers of new jobs had not been speculative. Unemployment among males in Derry in 1991 was the second highest in the north. Long-term unemployment was increasing. When Jim Swindall, of the Environmental Science and Technology Research Centre in Queens University, Belfast, told Radio Foyle listeners that the incinerator would attract new industry and new jobs, Eamonn Deane, director of the Holywell Trust in Derry, responded with a letter. How was Swindall's centre funded, Deane asked, did it receive funding from the chemical industry and from the Northern Ireland government? And did it receive "funding, directly or indirectly, from the Du Pont company?"[26]

In his reply Swindall said he was "of the opinion" that "incineration is a highly efficient and effective method of destroying hazardous or toxic waste". He added that "everyone who has been objecting to the incinerator has been objecting to the wrong thing":

> They should be requesting that any incinerator that is built has state of the art monitoring equipment on it to raise the alarm if there is any release of harmful material at any time. The reason that incineration is such a controversial topic in Ireland stems largely from the incident at Clonmel with the Merck Sharp and Dohme incinerator. Apparently the equipment was not run at the correct temperature and toxic substances were released into the atmosphere. We should learn from the incident and not simply throw up our hands and say no to all incinerators.[27]

Swindall said that Du Pont ran "a superb incinerator at Beaumont in Texas" without problems. "Therefore," he said, "who better to run the incinerator in a responsible manner than Du Pont?" Swindall added that he had been contacted by the BBC "for comments on the proposal" and that he had "great pleasure in doing so" as he was "becoming increasingly impatient with the one-side, unscientific and incorrect statements being perpetuated in the media". At no time, he stated, had he been approached by Du Pont or by the Northern Ireland Department fo the Environment to comment on their behalf. "Personally, I believe that Ireland as a whole needs an incinerator and I would be much more happy if it were run by Du Pont at Maydown than some small relatively inexperienced organization in, perhaps, the West of Ireland."[28] He added:

> You may not accept that Ireland needs an incinerator but I have yet to see a viable alternative proposal from Greenpeace or anyone else. To suggest that the problem of toxic waste can be solved by not producing it is naive in the extreme. Responsible industry is making huge efforts to minimize toxic waste production but there will always be some produced which needs to be destroyed in an environmentally sound manner if we are to live in a modern society. Alternatives such as storage in drums and, more often, throwing it in holes in the ground and hoping it will go away are totally unacceptable as is exporting it to some other country for incineration. I believe that Du Pont have given an undertaking that if the feasibility study is positive and an incinerator is built then they will not accept waste from outside Ireland and this is to be encouraged.[29]

Swindall also claimed that "environmentally sound industry" would be attracted to Northern Ireland if they knew there was "proper provision for destroying the small amount of toxic waste all industry produces". In answer to Deane's questions Swindall said that the centre (known by the acronym QUESTOR) was funded by the International Fund for Ireland, industry and the Queen's University, Belfast. He added that the Department of the Environment was a sponsoring member of QUESTOR and that funding was accepted from the chemical industry, including Du Pont.[30]

Not everyone in Derry was reassured by the experts, as the letters pages in the local press continued to fill up with anti-toxic missives. One writer suggested that the "Yanks" should build their toxic waste plant "at 1600 Pennsylvania Avenue, Washington D.C. because they cannot poison Ireland. They say these plants are safe. I wonder where we have heard that before: you don't need an expert to tell you if you stick your hand in the fire you will get burned."[31] In the following issue of the *Derry Journal*, which reported that the two Irish environment ministers, Richard Needham and Padraig Flynn, had met and discussed Du Pont's plans during the Anglo-Irish conference in Dublin on Thursday 31 January, screaming headlines proclaimed: "No Danger Claim Totally Rejected"; "Massive Opposition to Toxic Waste Plan"; "Little Support for Toxic Waste Plant", and in the letters page the anti-toxic opposition was total.

The *Journal* revealed, as the result of its own survey, "that the vast majority of the Derry population are very much against the siting of a toxic waste disposal incinerator by Du Pont". Heather Diamond told the paper: "It is brilliant that it will bring jobs, but it is a definite no on my part because the risks by far outweigh the job factor." Paul Smith said: "Derry people are not stupid and they need more than verbal assurances from the company about safety. In fairness Du Pont have brought much needed employment to the area. But that is not the point. Today they may only dispose of Ireland's toxic waste. But what is to stop them from tendering their incinerator services to other countries tomorrow." Hugh Hegarty said the authorities "should take steps to clean up the environment instead of adding to the pollution already present".[32]

Swindall's comments, and the news that Flynn and Needham had met, brought responses from the community. The Muff correspondent for the *Journal* was scathing in his criticism of Swindall. "Mr Swindall may find that the natives in this part of the woods are not as green propaganda wise and (a lot greener environmentally) than they used to be and that although they live on its banks, to use an old Derry phrase, 'they didn't come up the Foyle in a bubble'."[33] Sinn Fein councillor Hugh Brady lambasted Needham. "Attempts by Richard Needham to link major job prospects in conjunction with a toxic waste incinerator are clearly seen by the people of this city as a feeble effort at dangling the economic carrot in front of a job starved people. Job starved we may be, soft on the environment he may believe, but soft in the head we most certainly are not."[34]

Muff is the first small town on the coastal road north out of Derry into the Inishowen peninsula. Like Greencastle and Moville further up the road, these townlands are intricately linked to the fortunes and failures of the conurbation that has spread out from Derry city. Derry and Donegal people live and work in each other's parishes and counties. The concern in Inishowen about Du Pont's proposed incinerator was based on a knowledge of pollution among the environmentalists, and a fear among local politicians that something was happening over which they had no control. "Two-thirds of the time the prevailing winds come from the south/south-west," said the campaigner Enda Craig. "That would mean that toxic emissions from an incinerator in Du Pont would land on the Inishowen peninsula. Greencastle and Moville are in the slipstream of Du Pont."[35] Donegal County councillors representing the Inishowen natives were concerned because all their knowledge about Du Pont's plans came from media reports. A discussion in Donegal County Council was essential, councillors McGuinness, Fullerton and McGonigle asserted, to determine "the possible effects on County Donegal and particularly the Inishowen peninsula". Councillor Bernard

McGuinness also wanted to know whether it was necessary to employ independent experts to assess Du Pont's plans.[36]

While the committed environmentalists and community activists were organizing public meetings, lobbying local politicians and supplying the media with information, the ordinary person in the streets and villages in counties Derry and Donegal worried about the consequences of the protest. Community worker Eddie Kerr, in his research for a BBC Ulster programme on the issue and in his profession, had heard most of the arguments and listened to the fears:

> Although it was quite evident that the vast majority of people were against the siting of such a process in their area they were unsure why. Public reaction was muted in the face of a barrage of chemical compounds that blinded the issue with science. The language of protest was filtered with abbreviations that couldn't be elaborated upon and repeated doomsday scenarios of Bhopal and Seveso. As the days went by the campaign was tempered with the fear of offending Du Pont without whom the city would slip into the abyss of total economic obscurity. Du Pont was, several public meetings were constantly reminded, "our biggest and best employer" who would not take it too well if the city turned against it.[37]

Throughout February the issues became clearer, as did the legacy Du Pont were leaving the people of Derry and surrounding areas.[38] It also became clear to those opposing Du Pont's plans that they were now fighting three separate campaigns: one to stop the US multinational establishing a "national" toxic waste incinerator in Derry; the second to discover more about the company's application for a solid waste burner which would incinerate lycra waste; and a third to investigate the company's past methods of disposing of their toxic and hazardous waste, whether this and other processes had caused not only local environmental damage but also harm to workers and the Derry and Donegal population. But it would not be until after the first anti-toxic rally in the Guildhall Square in Derry on 16 March that these issues would become manifest within the campaign. Du Pont's proposals for a "national" toxic waste incinerator had to be stopped first.

Du Pont had insisted all along that there was no link between their application for a lycra waste burner and their proposals for the "national" incinerator. On 10 February, a report in the *Sunday Business Post*, which quoted a Northern Ireland Department of the Environment spokesman, appeared to confirm this. Du Pont's application, submitted to the local authorities in December, was for a replacement solid waste burner and nothing else, said the spokesman. If Du Pont wanted to build and operate a "national" toxic waste incinerator they would have to submit a separate application and an environmental impact study. There was definitely no link between the two, he said. "We will want to know what we are being asked to give planning permission for."[39]

Some campaigners, who had been doing their sums based on information about Du Pont's processes, were not convinced. Du Pont had admitted in their January "Facts" document, that approximately 700 tonnes of hazardous waste were sent to Finland and France each year for incineration.[40] The Rossi report into "environmental issues in Northern Ireland" reported that Du Pont's waste was in fact shipped to Finland and that it was waste "from fibre manufacture". The Rossi report also revealed that Du Pont's contract with the Finnish incineration company and Finnish government was on a monthly basis.[41] It was clear to the campaigners that, with 1992 around the corner, Du Pont had to take steps to deal with their own waste. Yet, some campaigners wondered, if Du Pont build a "lycra waste burner" to dispose of the 700 tonnes sent each year to Finland and France why did the company need another incinerator? They also wondered about Du Pont's comment that "the type of incinerator being considered needs to be operated with a continuous feedstock. Maydown doesn't generate enough of its own waste to maintain a constant operation".[42] Did this mean that the new "lycra burner" and the Chemical Waste Incinerator, which had been operating in the Maydown complex since 1981, would become obsolete if Du Pont went ahead with its plans to build a "national" incinerator? The "lack of any real understanding and information about the nature of the proposed waste incinerator", which Eddie Kerr had talked about, was a significant factor.[43]

In the meantime the SDLP's John Hume was still asking questions in Westminster. The plot thickened when Hume asked Needham if any Northern Ireland government agency had communicated with Padraig Flynn before the latter's November 1990 correspondence. Needham said: "No."[44]

Hume received this answer on 11 February. Five days earlier Flynn had responded to questions in the Dail (the Irish parliament) about Du Pont's "national" incinerator. Flynn confirmed that he had held discussions with Needham during the Anglo-Irish conference on 31 January on "the likelihood of the whole of Ireland being able economically to support more than one facility" for hazardous waste incineration. Flynn reiterated what he had been saying since the beginning of the year, that he intended "to allow time for the situation to clarify further before deciding to support any particular proposal".[45] What he did not reveal was the important fact that it was his department which had responded to Aspinwall's suggestion, not the north's Department of the Environment. Needham later told Hume, on 18 February, that no "agreement or understanding" had been reached between himself and Flynn regarding Du Pont's proposals. Needham added that "a major development involving a hazardous waste incinerator would not be determined without a public inquiry".[46] But there was still the small matter of Du Pont's application for a "lycra" waste burner, and the EIS which would go with it. Nobody had asked specific questions about it or about whether there would be a public inquiry.

John Hume's parliamentary stance (the Foyle MP had tabled 15 questions in 22 days) did not impress the campaigners. The information he had procured from Hanley and Needham told the campaigners little they did not know from their own studies of regional and EC legislation and from their own investigations. Hume had still not spoken publicly about Du Pont's proposals. The campaigners wondered why and decided to find out. A press release from the DDEC was dispatched to the *Derry Journal* for publication in the issue of 26 February. The headline was to the point: "Toxic waste: Where does John Hume stand?"

> We would like to pose a direct question to John Hume on this matter. Will he state publicly where he stands on the proposed siting of a national toxic waste incinerator in Derry? Is he or is he not opposed to this proposal? As an elected representative he should reflect the views of his constituents, and public opinion so far has come out very strongly against their proposal.

A spokesman for Hume responded in the following issue of the *Journal*. "Having campaigned in 1984–85 to get toxic waste dumping in the Atlantic by Britain stopped, John Hume is hardly likely to be supporting the importation to Derry of toxic waste from anywhere else"[48] The following week, after Hume had met with some of the campaigners, an SDLP spokesman said that the DDEC had "got the wrong end of the stick" and the purpose of the meeting was to clarify the position. Hume agreed with the campaigners that there should be no toxic waste incineration in Derry, no importation of toxic waste into Derry, and that Du Pont should put a greater effort into exploring methods of production which would reduce or eliminate the production of toxic waste.[49]

As the first anti-toxic rally (organized by the DDEC) neared, opposition to Du Pont's proposals spread into the heart of Donegal and into the towns around Derry city. The chairman of the North-East Donegal

Fine Gael constituency executive said that the majority of people in the country were "united to a person" in their opposition;[50] in Limavady, during a council debate on the issue, the "general consensus" was against the proposal. One councillor said: "Du Pont are one of the biggest firms around here. We want the jobs but not the waste. Let them throw it somewhere else." The chairman said: "I think we should wait until the feasibility study is ready."[51]

The feasibility study and the rally were still future events when Du Pont called their first press conference on the debate, for Tuesday morning, 5 March. Maydown site manager, Peter McKie, said Du Pont's lack of "input into the public discussion" about the proposed incinerator was "due to circumstances which were beyond our control" but he added that the public comment was "uninformed" though "well intentioned". If we build the incinerator, he said, "it will be to the highest possible specifications known at that point in time . . . by using state of the art technology it can be turned into something no more environmentally difficult than a number of coal fires".[52]

The local and national media in Belfast and Dublin faithfully covered the conference and quoted McKie extensively. The campaigners wondered who Dr Ruth Patrick was, described by Du Pont as a "world leading environmentalist" and who had carried out a survey on the river Foyle in 1959, before operations had begun at Maydown. This survey, said Du Pont, had been repeated in 1967. Du Pont added: "BEtween 1971 and 1989 the number of salmon passing up the river increased. Mussels in the region have shown no detrimental effects. Soil samples taken from the area in 1988 compare favourably with rural areas from all over the UK. No dioxins above the naturally occurring level have been traced in the soil."[53]

This information, several campaigners believed, was irrelevant to the debate. Given the fears about pollution in the Foyle, would anyone actually dare eat the mussels, which are prime feeders and would not show detrimental effects from pollution anyway; the fact that there were more salmon in the Foyle did not mean that the fish were not poisoned. No one had mentioned the existence of dioxins in the soil around Du Pont; it was around the ReChem incinerator in south Wales that dioxins had been found in the soil. The Derry and Donegal communities simply did not want to find dioxins on their own doorsteps.

Despite the concerted opposition (approximately fifty groups had been formed) Du Pont seemed to believe they could convince the people of Donegal, Derry and the surrounding townlands that incineration was safe. "There has to be some means of safe disposal," a spokesman for the company said during the week before the rally. "We believe incineration is the best way of doing that." It wasn't the view shared by the campaigners, who dismissed McKie's coal fires analogy. "Du Pont are insulting the people of Derry and insulting their intelligence with comments like that," said the DDEC's Jim O'Neill.[54]

The morning of the rally was overcast, the rain heavy; it was a day for staying in, but it was also Saturday, shopping day in Derry and St Patrick's bank holiday weekend. Greenpeace had brought a bus and some supporters up from Dublin. Ralph Ryder, from Ellesmere Port (see Chapter 4), Fiona Sinclair, from Inverness (see Chapter 2) and Clare O'Grady Walshe, of Greenpeace Ireland joined local speakers Enda Craig, Eamonn Deane and Jim O'Neill as the crowds began to gather in the Guildhall Square. A few minutes after three o'clock, Deane stepped up to introduce the theme for the day: "Toxic Waste – No thanks". Throughout the afternoon the colourful anti-toxic banners contrasted with the dull skies; crowds drifted in and out of the square to hear the speakers, with an estimated 5,000 passing through during the two and a half hours of the rally. Eamonn Deane's address provided the refrain for the afternoon:

We are for a better quality of life for our children and our children's children; for a cleaner and safer environment; for advances in technology which enhance the environment; for creating meaningful and useful jobs; for greater access to information and information systems and empowering ordinary people to decide what happens in their community, instead of letting the men in the grey suits decide. Birds of Passage politicians in Belfast or Dublin will not decide how we live.

It was reiterated by all the speakers: "The loud and strong message going out here today is that people in the North West and Ireland don't want a toxic waste incinerator," Jim O'Neill, the final speaker, concluded.[55]

It had been a peaceful celebration of people power. The local (and to a lesser extent) the national media, accustomed to mass civil rights, republican and unionist rallies, reported the crowd as between 1,000 and 2,000. One reporter contrived to report that the organizers were disappointed with the turnout. With groups pouring into Derry from all over the country the true disappointment, according to several campaigners, was that not enough Derry "working class or unemployed" people had given their active support. For a city previously without environmental awareness it was, several groups insisted, an overwhelming success. Environmental and community groups and activists from Cork, Dublin, Belfast and all the major towns in the North West mingled with residents, with school groups from Derry and with the myriad of environmental groups which have formed in Derry and Donegal since the campaign began. Only the media, it appeared, were disappointed with the attendance.

The story that Du Pont had been fined $1.85 million for a hazardous waste violation at its New Jersey works broke the following day in a two paragraph piece in the *Sunday Business Post*.[56] The campaigners had learned about the fine some days earlier. It was, for many campaigners, the perfect riposte to Du Pont and the company's apologists' claim that the US multinational "would not engage in any project that is harmful to its employees, the neighbouring environment or local community".[57] If they could make mistakes in the US, then surely they were capable of making mistakes in Derry. Du Pont's increasingly frequent forays into the debate and statements to the media were seen as pure propaganda, particularly when the company boasted that their "track record, undisputed technical expertise, total commitment to safety and environmental concern, make it the ideal overseer of such a project, given that an incinerator is necessary".[58] The campaigners disagreed: "Given Du Pont's record, we wonder whether they should be entrusted with the safe running of a hazardous waste incinerator, if such safe running is indeed possible."[59]

Du Pont were not ready to give up their plans; the feasibility study, Du Pont's Ray Bradley said, would be ready at the end of April. The company had provided a telephone information service; a fact sheet was distributed to approximately 40,000 homes, one of the biggest industrial information campaigns ever launched in Northern Ireland. The Derry public were told that incineration was safe. "Emotive language can be misleading," Du Pont chemist Dr Patricia Shaw said. "If people concentrate on all the facts I think they'll be reassured." The Du Pont document "The Facts" had brought messages of appreciation from the Derry public, said Bradley:

> The community is becoming increasingly aware of the balanced response that Du Pont is making in the face of misunderstandings and in some cases misleading information about a proposed incinerator. While the decision to proceed has yet to be made, many key points are now widely understood. No waste will travel up the River Foyle. Waste which would only come from the rest of Ireland and would constitute only one or two lorry loads, in sealed tankers or drums, per day. It would definitely not include PCBs, radioactive material, clinical or animal waste.

He said that the public had appreciated there would be a "totally independent" EIS (Environmental Impact Study), which would be made available before construction commenced.[60]

This did not placate the communities in Derry and Donegal, who decided to do their own environmental impact study. Peter MacKenzie of the DDEC said:

> Due to the failure of the DoE to effectively monitor Du Pont's toxic waste or any toxic waste producers, we feel, as a community, we have to fully investigate the implications of an incinerator being located in our area. Such an investigation will include an assessment not only of the environmental hazards posed by normal or accidental working of such a plant but also will assess the likely impact on farming, fishing, tourism, the health of the population and the very serious threat to loss of employment in all of these areas.[62]

In a press statement sent from the DDEC on behalf of the campaigning groups Du Pont were left in no doubt about the level of sophistication amongst their opposition:

> We have energy consultants, engineers, marine biologists, farming experts, designers, architects and health workers actively engaged in the campaign, not to mention a whole host of people with that great common asset, common sense. We do not intend to see our future flushed down a toxic toilet by some dry report from an academic department.

On the same day that the local media reported the new initiative by the campaigning groups the Campsie residents called on Du Pont to "come out and talk to their neighbours".[62]

Any evidence of a dirty tricks campaign by the company, its apologists or by the campaigners was not apparent in the aftermath of the rally, but some curious correspondence managed to creep into the local press. One letter writer suggested that Sinn Fein had been "hijacking everything" and had been too much in evidence at the rally. Eamonn Deane, of the Holywell Trust and one of the rally organizers, responded: "The Toxic Waste – No Thanks Campaign is rigorously non-party political. No political party speakers addressed the rally and no politicans were on the platform. The campaign unites people from all walks of life around a single issue – their opposition to Du Pont's plan to build a toxic waste incinerator."[63] Peter MacKenzie, in an individual capacity, responded to another letter, also in the *Derry*

Journal of 22 March 1991, that the campaigners were "professional protestors". MacKenzie said:

> Unlike the USA we have no right of access to information, nor do we have a regulatory body that has the staff and resources to adequately monitor toxic polluters. So if we are to make an assessment of Du Pont's safety record we have to look at the company's performance in the USA:
>
> 1.   Du Pont is top of America's top ten toxic polluters. According to the government's Environmental Protection Agency's figures, Du Pont released 151,078 tonnes of toxic waste (in 1990).
> 2.   Du Pont is the world's greatest manufacturer of CFCs (it has 50 per cent market share). Although there has been scientific evidence that CFCs have been destroying the ozone layer for the past 17 years, Du Pont has resisted any attempt to stop production of CFCs. Public concern about CFCs in aerosols led to a boycott of aerosols and there was a dramatic drop in the demand for CFCs. With its renowned concern for the environment Du Pont embarked on a massive capital investment programme to find alternative uses for CFCs. It found such a use in the computer industry.
> 3.   Du Pont's Chamber Works in Deepwater, New Jersey has so polluted the groundwater that it has agreed to pay $39 million to decontaminate the groundwater. It is in the same works that Du Pont have recently been fined $1.85 million for repeated violations of its discharge limits. Four officials from the plant have been convicted for concealing asbestos hazards from employees suffering from asbestosis."

MacKenzie went on to look at the history of Du Pont's Maydown works where, he said, the company had been operating a toxic waste incinerator without a licence for a number of years, and had been dumping hazardous waste into an enlarged landfill originally designed for construction waste. Why, he queried, did Du Pont use subcontractors to handle its waste disposal if it had the "scientific skills and integrity" to make Derry "Ireland's toxic dustbin".[65]

Du Pont responded immediately in the following issue of the *Derry Journal*. Ray Bradley said Peter MacKenzie had made "several untrue and misleading statements" about the company's operations at Maydown. (Nothing was said about the US allegations.) "The incinerator [in Derry]," Bradley said, "is and always has been properly authorized by the Alkali Inspectorate". The landfill, he said, had not been enlarged; "the site landfill was properly permitted by the City council". Bradley added that all hazardous waste operations were carried out by trained Du Pont personnel; "this includes the loading of containers". Waste shipped to Finland for incineration was handled by the agent for the disposal company who, he said, "comes on site to carry out an independent check on the loading of the container and the paperwork prior to shipment".[65]

Bradley's statement about the incinerator was, in fact, correct. Derry City Council did not know about the "chemical waste burner" because they did not *need* to know. When Du Pont started up the burner in 1981 they went to the DoE in Belfast for planning permission. "No licence was required," said a DoE spokesman, "because it is an integral part of the process and Du Pont had a specific exemption". The Alkali Inspectorate, he added, were aware of the burner.[66]

The Campsie residents had already been told this when Bradley responded to MacKenzie in the newspaper. In what some campaigners saw as a surprise initiative by Du Pont, the company invited CRATE to the factory on Wednesday 3 April. During a four-hour meeting Du Pont said that they would keep the small chemical waste burner going, but still wanted permission to operate the new lyrca burner. The capacity of the new burner would be 1,000 tonnes per year and would burn lyrca waste only. Did this mean, several campaigners wondered, that Du Pont did not plan to go ahead with their proposals for a "national" incinerator?

Letters sent out by Peter MacKenzie on behalf of the DDEC to the clergy, the council, John Hume and Richard Needham about whether the replacement lycra burner would be subject to a public enquiry as well as the national incinerator brought some swift responses. The Bishop of Derry said he had been carefully reading all the material published in the local media. He said he was concerned about the proposal and would only make a public statement or enter the debate when he had all the facts. The Bishop of Derry and Raphoe made the same comment.

MacKenzie had asked the recipients of the letters to indicate whether they wanted additional information on the environmental impact, health effects, efficiency of incineration, alternatives to incineration and Du Pont's safety record. Several councillors responded immediately but not everyone liked the tone of the letter that asked them to state their position on the proposed national incinerator.

The leader of the Ulster Democratic Party and Derry city councillor, Ken Kerr, said that his main concern was "that we are not used to get rid of the Republic's problems". He questioned the motives of those involved in certain campaigning groups, claiming that they seemed to be more concerned with throwing mud at Du Pont than with putting the facts before the public.

Any statement I have (made), or will, make on the issue will be based on balanced and unbiased checking of information that has been put before me. May I first suggest to those that claim to speak on behalf of people or groups that if they believe that they have the support of the people that they put themselves put for election at the next council elections and if elected can then truly claim to represent the people."[67]

Eamonn Deane and Peter MacKenzie immediately responded to Kerr's attack on the campaign with a press release which went to the Derry media and to the *Telegraph, Newsletter* and *Irish News* in Belfast. The authors said the campaign was opposed to the national incinerator because:

There are real and genuine fears about a toxic waste incinerator;
    The transportation of toxic waste to this proposed incinerator is opposed by all public representatives including (we are pleased to note), Councillor Kerr;
    It is a reasonable expectation that the incinerator would attract other "dirty" industries;
    We believe that it is of vital importance that the views of people here be heard on this issue – not just the views of elected representatives;
    We do not believe the DoE currently have the capability to monitor a national toxic waste incinerator;
    Since the process of high temperature incineration results in the release of highly toxic substances such as dioxins, furans and other chemicals previously unknown it is out contention that the only way to determine the health implications of incineration is by a national survey of health in communities that currently have incinerators.

Deane and MacKenzie said that the information used in the campaign had been "drawn from official government reports, reports commissioned by government bodies and academic research conducted in University research departments". It was, they added:

insulting to the people involved in the campaign and to the community at large to suggest we cannot read and understand such reports. One thing is for sure, the people of Derry have been told too often in the past to stay quiet and that somebody else knows what is good for them. We urge everyone to express their feelings on these issues whether they are in favour or against the proposed incinerator".[68]

The campaign had extended beyond Derry and Donegal during March, at the height of the rally. Many of the newer groups, among them the

Magilligan Area Residents Action Group and Limavady Against Toxic Emissions (LATE), had taken Deane's and MacKenzie's advice. Eric Gingles of the Magilligan group said that it was their campaign strategy to collect and collate information on Du Pont worldwide "to counteract the environmental rhetoric they have propagated throughout the North-west" and to investigate the "track records" of both the Alkali and Radiochemical Inspectorate of the DoE and the Health and Safety Inspec-torate of the Department of Economic Development *vis a vis* Du Pont.[69]

LATE had collected more than a thousand signatures from people opposed to Du Pont's proposals when Henry Reid of the Ulster Farmers Union told a public meeting in the Limavady Town Hall on 11 April 1991 that his members in the area had unanimously opposed the incinerator. He said the land would be destroyed if Du Pont built the incinerator. A woman said she would buy imported meat and vegetables, which would mean a loss to farmers in the area and a loss of jobs. Mick Conlon of the DDEC said the rally in March had legitimized the campaign. "The protest was right; it was the people having a voice." Paul Fleck, chairman of LATE, said it was not the aim of the meeting to blacken the name of Du Pont or force them to close down. "We want Du Pont to reconsider its proposal to site an incinerator at Maydown which will affect the whole of the north-west, including Limavady."

As April drew to a close there was still no sign of Du Pont's feasibility study, but the company admitted that preliminary results from the study showed it was "technically feasible but not financially viable" unless government met part of the cost.[70] The authorities in Belfast had refused to comment on whether Du Pont would be entitled to grant aid for the project. (Total investment by Du Pont in Derry has been estimated by the company at over £600 mil-lion – over half of this since 1980. Grant aid to the company has been as high as 50 per cent for specific projects. In 1989 Du Pont received from the IDB a capital grant of £4.7 million for assistance in employment creation and renewal of jobs. That year Du Pont's worldwide profits were $2.4 billion. The Maydown operation would have contributed between four and five per cent of those profits. If Du Pont were allowed to build a national incinerator it would be able to reduce costs, increase productivity and increase its profits. It would also make the Maydown complex virtually self-sufficient.[71] Whether the incinerator would be financially viable would depend on the availability of government grants and whether contracts for the disposal of toxic waste from the Republic of Ireland could be secured.) Du Pont also admitted that the incinerator, if built, would create no more than 60 permanent jobs by 1995.[72] Peter MacKenzie responded:

> We have repeatedly said that there is insufficient toxic waste produced in this country to justify such a facility and that it would be a much better strategy to support a toxic waste minimization programme, full environmental audits on all toxic producers and research into the alternatives to incineration.[73] This cannot possibly be deemed good value for money in terms of job creation, coupled with the fact that this will cost an untold number of jobs in farming, fishing and tourism.[74]

Despite internal differences between some of the campaigning groups over strategy and timing, the protest was carried into the council chambers where a deputation from the Moville/Greencastle group, the Ulster Farmers Union, CRATE and Muff Irish Countrywoman's Association addressed Derry's councillors. They put specific proposals to the council to reject the incinerator and the importation of waste into Derry, and to initiate an enquiry into Du Pont's present toxic policy. Before the council unanimously adopted the proposals in principle, Dan McGuinness of the Moville group said there was an increasing conflict between industries which produce toxic waste and the communities which host them. The idea of a national incinerator and the selection of a location for it really meant that the government accepted, on behalf of the people, responsibility for corporate toxic waste dumping and selected the local community who would take the consequences. Sinn Fein councillor Hugh Brady proposed that the council accept the deputation's proposals in principle and refer them to the Environment Protection Committee.[75]

Meanwhile the DDEC, on behalf of the remaining campaign groups, had prepared a similar set of proposals to be put to the council. The last three proposals stated:

> We are concerned that there is not sufficient information presently available to ascertain the safety implications of the operation of a toxic waste incinerator. We are convinced that a full epidemiological survey of the health of communities wherein incinerators are located is the only way to determine the possible implications of the siting of an incinerator here.
>
> We feel that an inadequate case has been made for a national toxic waste incinerator. So far the only criteria applied have been those of economic and commercial viability. Without full environmental audits

being done by major toxic waste producers we cannot say what volume of toxic waste needs disposed of.

We will adopt a policy of promoting the public right to information so that councillors and the community will have access to the information necessary to evaluate proposals put forward by businesses seeking to operate in the North-West.[76]

The campaigning groups had been busy in the period after the rally. Rosemary Vaughan and Peter MacKenzie compiled a 29-page briefing docment on the proposed incinerator for the DDEC. They examined four specific areas:

● Why a national toxic waste incinerator is being planned;
● Legislation, regulation and monitoring;
● What do we know about incineration; and
● Du Pont International's technology.

Until this document appeared, few of the campaigning groups had expressed their feelings about the existing legislation, which the majority of the community believed would protect them. "Many people feel that the proposed incinerator will be OK simply because they believe that if there were any problems then the DoE would not allow it to go ahead. We have been told by certain elected representatives that any facility would be of the highest standards," they wrote. The problem with this approach was:

● Who sets the standards and how are they set?;
● Who monitors the industries compliance to these standards?; and
● How effectively are breaches of these standards dealt with?

There is a long history of attempting to set safety standards for a wide variety of modern activities. The primary difficulty is trying to reconcile to opposing forces maximization of safety and maximization of profit. Clearly most safety procedures cost money either directly in capital costs or indirectly in slowing the process or requiring greater staff time.

The problems for setting standards in environmental health are even more complex. Because it is difficult to prove the direct causal link between a particular activity and some health effect, deciding on what represents a significant danger is often a point of controversy. Our knowledge of the impact of environmental pollutants is growing all the time but we have only begun to look at the problem seriously in recent decades. ... We know that it is difficult to establish safe levels of emissions because there is no such thing as absolutely safe and we know that there is a tension between safety and cost, so how are standards set? ... The underlying assumption is that these processes MUST take place so we must settle for WHAT INDUSTRY IS PREPARED TO PAY FOR. One of the most powerful forces for change in improving safety standards is public opinion; unfortunately most people are unaware of the nature of the hazards from toxic emissions. Unlike the US we have no public right to know, we cannot ask those industries in our area what they are burning and why. Information on leaks and accidents is almost impossible to obtain. Industry regularly covers up its accidents or buries them in long technical reports instead of clearly admitting the problems. Until communities have a legal right to information that directly affects them we cannot hope to see significant improvements in our environment.

MacKenzie and Vaughan choose not to record the history of governmental control of pollution, "but put simply," they stated;

there have been a variety of governmental agencies attempting in piecemeal fashion and without adequate resources or personnel to monitor the worst excesses of industry. Much of the monitoring that is done is self-monitoring by industry and where any governmental agency intended to actually call on a particular business they did so by appointment. It was not possible therefore to identify particular routines that regularly made a mockery of sampling or monitoring. Until the monitoring of pollution comes under the strict control of one statutory body (which is in itself fully publicy accountable) with adequate resources to implement a proper programme of monitoring we cannot expect what regulations we do have to be adhered to.

It may seem ridiculous to have to say this but there is no point in having regulations about emissions if no action is taken when those rules are flouted. For the ordinary mortal our lives can seem bound by a mountain of petty rules that have dire consequences if we do not comply, not so for large business. It is absolutely commonplace for breaches of emissions limits to go completely unchecked. There seems to be a gentleman's agreement that "oh well you have a difficult job to do and you are doing your best". In this context there is little reason for industry to take environmental control seriously.[77]

It was clear to everyone except Du Pont, it seemed, that the incinerator was not wanted in the north-west. The *Derry Journal*, in its editorial of Friday 3 May 1991, was behind the people. There were good reasons, the leader writer stated, for the opposition. "This city has suffered enough from deliberate discrimination from Stormont governments over two generations. The people are in no mood to allow the installation of a plant they believe would damage its environmental attractiveness and stunt hopes of economic revival." The people of

Inishowen, said the writer, had not been treated well by Fianna
Fail governments despite the loyal support for Fianna Fail down the
years. The writer was critical that Padraig Flynn had refused to meet
Donegal represenatives to discuss the Irish government's attitude on
the incinerator. "There should be renewed efforts from Donegal to
get the minister to come clean on what he discussed during his visit
to Derry. The threat of the loss of a traditional seat in Inishowen
might wonderfully concentrate government minds in Dublin." The
editorial was headlined "Derry Says No" and the writer concluded:
"The dismissal of community protests from Derry played a major
part in precipitating the civil rights movement in the North. The toxic
incinerator has become the biggest issue to command support across
political divides since those times. Both the Northern Ireland Office and
Dublin should tread warily."[78]

Donegal county councillor, Bernard McGuinness, was in no doubt
about Flynn's treatment of the people of Donegal, especially those in
Inishowen. "The minister is quite obviously going to consider a toxic
waste incinerator for the whole of Ireland if the feasibility study shows
that it would be viable. . . . By refusing to meet the Council on this matter and not giving enough money for housing, the Minister is treating
us with contempt. These issues will be vital in the local elections."[79]

As the marching season in the north of Ireland drew closer there was
still no sign of Du Pont's feasibility study, which had been promised
within a few weeks at the end of April. After the abrupt hiccup over
tactics in April the campaign was united again in May when the
anti-incinerator groups agreed in principle to three of the following
objectives:

1. The campaign groups are opposed to the siting of a national toxic
   waste incinerator in the north-west of Ireland;
2. The campaign groups are opposed to the siting of any new hazardous waste incineration facilities in the north west of Ireland;
3. The campaign groups are opposed to all forms of hazardous
   waste incineration because they believe them to be an inherently
   dangerous and unsafe technology; and
4. That groups/individuals remain autonomous and act on their own,
   if not in agreement with the principles established above.[80]

On the day the groups were expected to pass the resolutions the
Derry Journal published a column by Peter MacKenzie titled "Why you
should worry about a toxic incinerator". MacKenzie said: "Although
there has been a number of articles in the press I can't help feeling

that many people still are unsure of what the issues are", adding that the opposition to the proposal was based on the belief that "incineration is both an unknown and unsafe technology". He said he had started from a position of ignorance and at one stage "was almost beginning to believe the information being pumped out by Du Pont, but fortunately I kept on researching the whole question of incineration and now I am convinced that if this incinerator was to be built it would be a disaster for Derry". He said the main facts that needed to be considered were:

1. High temperature incineration creates a multitude of new chemical compounds; less than 10 per cent of the chemical compounds coming out of the incinerator stacks have been identified. The toxicity of the other 90 per cent is UNKNOWN.
2. The much quoted destruction removal efficiency of 99.9999 per cent refers to only 1 per cent of the chemicals coming out of the stacks and refers to ideal test burn conditions. In the commercial environment it is unlikely that such figures will be achieved.
3. The burning of solid toxic waste leaves behind 25–35 per cent of its volume as highly toxic waste which will have to be buried here in the north-west.
4. The reality of toxic incineration is that experienced by Mr Hanrahan in Clonmel, whose whole herd of prize animals were destroyed and whose family's health was irreparably damaged by an incinerator run, not by "cowboys", but by one of the world's leading pharmaceutical companies.
5. Heavy metals cannot be destroyed by incineration but are pumped into the environment.
6. As more and more countries are beginning to realise that their whole agricultural system is so polluted by industrial residue they are keen to buy farm produce from Ireland because of its green image. This marketing advantage will be lost if a national incinerator is sited here. The fishing community, too, would suffer a severe economic penalty.
7. Doctors in Germany are advising women to restrict their breast feeding of babies to three months because the levels of dioxin in their milk is dangerous. We must begin to take seriously environmental pollution when a mother cannot feed her children.

He said: "By continuing to build newer and bigger incinerators we are simply letting toxic waste producers off the hook, they have no incentive to reduce the toxic waste stream." The solutions for industry and government he outlined in five points:

1. Like the recent decision of the Canadian government, we should commit ourselves to a policy of building no new hazardous waste incinerators.
2. We should insist on much greater independent monitoring of toxic waste producers.
3. We need more research done on the environmental and health implications of current incinerators.
4. All toxic waste producers should be forced by law to submit themselves to a full environmental audit; to publicly release information on the nature and volume of all its toxic waste streams, in particular the mass balance data – that is the only effective method of estimating the toxic pollution.
5. All toxic waste producers should be made to submit detailed plans of their waste reduction programme. If the claim by Du Pont that they intend to reduce their toxic waste by 75 per cent is true, then they will not mind explaining how they intend to achieve this.

The single most difficult thing to deal with in this campaign has been the reluctance of individuals and public representatives to criticize Du Pont. It is, after all, the single biggest employer in the area. This is a very real concern and one I do not take lightly. Like any multinational company Du Pont's loyality lies with its shareholders and their profits, and as such its policies are determined by that fact. The company makes a £20 million profit from its activities here and this profit will not be substantially reduced if the national incinerator is not built. Of course, they could always move elsewhere, but incineration is being opposed all over the world and the capital lost in such a move would not be justified.[81]

However, the groups could not agree about the specific aims of the campaign – whether they were opposed solely to the national incinerator, whether they were opposed to incineration *per se*, or whether they were opposed to Du Pont's waste disposal methods. What had been a model campaign with a mixture of initiatives based on the experiences of other anti-toxic campaigns, and on the experience they had gained throughout the year, had begun to fall apart, and then become political. When several Derry City councillors spoke on a platform at a community-organized anti-toxic rally some campaigners wondered what had happened to the non party-political stance.

The management in Du Pont knew nothing about the internal squabbles amongst the anti-toxic campaigners. The company had their own problems to solve; the immediate one was how to convince the communities in and around Derry that incineration was safe. On Tuesday 20 August, Du Pont announced that they would begin an environmental impact study (EIS) as a precursor to their planning application for the national toxic waste incinerator. They immediately refuted claims by Peter McKenzie that they would have to import waste to make the project viable.[82] Richard Wilie, the Du Pont chemical engineer who had been given responsibility for the incineration project, said that the project was commercially viable[83] and argued that the EIS would state where the waste was from. He added that they had not considered the possibility of importing waste from other countries.[84]

During late August, in the wake of Du Pont's announcement, speculation about the funding of the project was rife. Du Pont had estimated that the incinerator would cost them £40 million.[85] The campaigners believed that the company would expect at least 65 per cent in grant aid, but Wiley would not reveal the percentage of grant aid Du Pont expected to receive from the relevant authorities. The speculation was that the aid would come from the EC Regional Fund via the Industrial Development Board, and possibly from the Department of the Environment in Dublin. In August a spokesman for the IDB said they were still in discussion with Du Pont. A spokeswoman for the DoE in Dublin said any application for aid from Du Pont would have to go through the relevant authorities in Belfast who would then pass it to Dublin on behalf of Du Pont.[86]

Du Pont expected to receive a "sizeable chunk of grant aid", as Wiley put it. The campaigners believed that the company would still have to invest at least £14 million in the project, and they argued that even if Du Pont received all the toxic waste currently generated in Ireland, as estimated by consultants Byrne O'Cleirigh, they would only return £2.5 million a year on 1991 disposal costs. "They are unlikely to be engaging in major capital investment without being certain of attracting revenue," campaigner Peter McKenzie said. "£2.5 million a year is not a good return on a £14 million investment, so if Du Pont are going to spend £14 million that would suggest they are going to import waste."[87] Wiley denied this and said that the incinerator would "not be a high profit making venture. That's our risk and we understand that". In 1991 Du Pont's annual generation of waste was estimated by Wiley at between 6,000 and 6,500 tonnes, which he claimed was not all toxic.[88] Increasing toxic waste disposal costs had pushed the price of disposal to £1,000 a ton (in 1986 it had been £300 a ton).[89]

With the campaigners unsure whether to believe Du Pont, Derry City Council agreed to ask Du Pont to provide a toxic release inventory, and to fund a health survey. Du Pont indicated that they would try to accommodate the council with information on toxic releases. "We would have to problem giving this information to the council and to representatives of the community groups," Wilie noted.[90] Under US legislation Du Pont are required to complete toxic release inventory forms for the Environmental Protection Agency. No such requirement exists in either British or Irish law. In February 1991 Irish Environment minister Mary Harney announced that she would like to introduce legislation which would require companies in Ireland to provide such an inventory.[91]

Concern about the health of people in Derry and Donegal had prompted Derry City Council to consider an epidemiological study. The council wrote to Bill McConnell, director of Public Health at the Western Health and Social Services in Omagh, to request information on epidemiological studies. Wilie said he was sure that Du Pont would welcome a health study but, he stressed, only when they knew what was involved. "We don't leave things to chance," he said. "No process is risk-free but we can build a facility to protect the public and give them confidence." There was no quick, easy solution, he said, but it was a priority at Du Pont to avoid, reduce and recycle waste and then find "an environmental way of disposing of the waste". Incineration, he added, was an environmentally safe method of waste disposal.[92]

It was clear to observers of the campaign that the two sides had reached an impasse. The community groups insisted that incineration was not a proven technology. The human health effects of incineration, they argued, had not been analysed and that even a full epidemiological study might not reveal meaningful results. Du Pont were convinced that incineration was safe.

Notes and references

1. For a comprehensive review of toxic waste disposal in Ireland see Allen, R. and Jones, T. *Guests of the Nation* pp. 208-246, 296-298 (Earthscan, 1990) and the Byrne O'Cleirigh report on the incineration of hazardous waste in Ireland (Byrne O'Cleirigh, 36 Upper Mount Street, Dublin 2, Ireland).
2. Toxic Waste Incineration Report, Derry City Council, March 1991
3. An Phoblacht/Republican News, 21 February 1991.
4. Bev Doherty is not entirely correct in her statement about temperatures though she may have been influenced by the Irish Supreme Court's ruling in the Hanrahan case. Justice Henchy, the ruling judge, said: "Because the function of the incinerator is to effect

the destruction by combustion of dangerous waste chemicals and solvents and because the incinerator was for significant periods in the years in question running at below its design temperature and therefore at a heat which was not adequate to destroy dangerous and contaminated solvents, it is marked out by the plaintiffs as the primary source of pollution on their farm." (See Mary, John and Selina Hanrahan v. Merck Sharp & Dohme (Ireland) Limited; 1982 no 2138P and 1985 no 316. Law Library, Dublin.) Further, no single authority has proved conclusively the precise pollution effects from incineration. Numerous reports and studies have been published on the incineration process and on the possible effects of pollution. If incineration cannot be proven to be safe then it must be unsafe. What is known is that all incinerators, working efficiently or not, create deadly poisons; it is the effect of these poisons that has not been conclusively proven, simply because scientists admit they do not know enough about the chemical compounds produced during the incineration process. While they have identified the toxicological properties of the majority of dioxins and furans, which can be created by the incineration process, only one, 2,3,7,8-tetrachlorodibenzo-para-dioxin (TCDD), has been properly studied yet there are 75 different types of dioxin and 135 furans. (See also Dioxins, furans, TCDD and 2,4,5-T, Chapter 9)

5. Doherty, B. "Toxic waste incinerator – no thanks" *Derry Journal*, 18 January 1991.
6. Ibid.
7. Ibid.
8. See *Belfast Telegraph*, 11 January 1991, *Irish Times*, 14 January 1991, *Derry Journal*, 15 January 1991.
9. Eire in waste disposal talks with Du Pont" *Belfast Telegraph*, 14 January 1991.
10. Allen, R. and Jones, T, pp 237-8, 242. Irish Environmental Services (IES) handle approximately 50 per cent of Ireland's toxic waste. SARP is a French waste management company.
11. Flynn, Padraig, speech given at Institute of Engineers of Ireland AGM, 11 October 1990, DoE, Custom House, Dublin 1.
12. Aspinwall & Co: "A review of waste disposal in Northern Ireland", for DoE (NI), June 1990, pp 57 (7.46).
13. *Belfast Telegraph*, 14 January 1991.
14. "Talks held in north on dumping of waste", *Irish Times*, 14 January 1991, *Derry Journal*, 15 January 1991.
15. For the full statement see *Derry Journal*, 18 January 1991.
16. *Irish Times*, 21 January 1991, *Belfast Telegraph*, 21/22 January 1991.
17. Waste incineration at Maydown – The Facts, Du Pont, Maydown, Derry (24 January 1991); see also *Derry Journal*, 29 January 1991 and *The Sentinel*, 30 January 1991.
18. Transcript of Radio Foyle news broadcast 31 January 1991 (morning).

19. Ibid.
20. House of Commons Official Report, columns 469-471, 29 January 1991.
21. Ibid.
22. Ibid.
23. Ibid.
24. Transcript of Radio Foyle news broadcast 31 January 1991 (morning).
25. Transcript of Radio Foyle news broadcast 31 January 1991 (lunch-time).
26. Deane, Eamonn, letter to Jim Swindall, 30 January 1991.
27. Swindall, Jim, letter to Eamonn Deane (Holywell Trust) 6 February 1991.
28. Ibid.
29. Ibid.
30. Ibid.
31. "No toxic waste Plant", letter to *Derry Journal*, 29 January 1991.
32. "Little support for toxic waste plant" *Derry Journal*, 1 February 1991.
33. *Derry Journal*, 1 February 1991.
34. Brady, Hugh, interview with Robert Allen. See also *Irish News*, 1 February 1991.
35. Craig, Enda, interview with Robert Allen. See also *Irish Press*, 28 January 1991.
36. McGuinness, Bernard, interview with Robert Allen. See also *Irish Press*, 28 January 1991.
37. Kerr, Eddie, "A burning question" Briefing document, March 1991 (unpublished).
38. There is now considerable concern in Derry about the number of Du Pont workers who have died from various cancers.
39. DoE (NI) spokesman, interview with Robert Allen. See also *Sunday Business Post*, 10 February 1991.
40. See note 18.
41. *Environmental Issues in Northern Ireland*, November 1990 (HMSO) pp 94 col 2.
42. See note 17.
43. Kerr.
44. House of Commons Official Report, col 330, 11 February 1991.
45. Dail reply to questions 24, 55, 74, 75, 150, 151, 152; 6 February 1991.
46. House of Commons Official Report, col 37, 18 February 1991.
47. Derry Development and Education Centre, press release, 25 January 1991. (All DDEC and campaign press releases and statements are on file at the DDEC office.) See also *Derry Journal*, 26 January 1991.
48. *Derry Journal*, 1 March 1991.
49. Durcan, Mark and O'Neill, Jim, interviews with Robert Allen 1991.
50. "Donegal united against toxic waste", *Derry Journal*, 1 March 1991.
51. "Councillors debate incinerator" *Northern Constitution*, 2 March 1991.
52. "Du Pont's proposed development", press release and documents,

Du Pont, Maydown, Derry 5 March 1991.

53. Ibid.
54. *Irish Press*, 16 March 1991.
55. For full coverage of the rally see *Derry Journal*, 19 and 22 March 1991, *The Sentinel*, 20 January 1991, the *Sunday Press*, *Sunday News* and *Sunday Life*, 17 March 1991.
56. *Sunday Business Post*, 17 March 1991.
57. See note 52.
58. Ibid.
59. "The hazards of hazardous waste disposal" *Derry Journal*, 29 March 1991.
60. See also *Irish Press*, 1 April 1991, *Derry Journal*, 2 April 1991, *The Sentinel* and *Belfast Telegraph*, 3 April 1991.
61. MacKenzie, Peter. Interview with Robert Allen 1991.
62. See *The Sentinel*, 3 April 1991.
63. See *Derry Journal*, "Disgusted" 22 March 1991 and "Non Party Political" 2 April 1991.
64. See *Derry Journal*, "The Du Pont Claims", 2 April 1991.
65. "Du Pont challenges misleading statements" *Derry Journal*, 9 April 1991.
66. Department of the Environment (NI), spokesman. Interview with Robert Allen 1991.
67. "Self appointed experts are alarmists" *Derry Journal*, 9 April 1991.
68. A full acccount of this statement is on file in the DDEC, 15 Pump Street and in the Holywell Trust, Shipquay St, Derry.
60. Gingles, E., correspondence with Robert Allen 1991.
70. *Belfast Telegraph*, 17 April 1991.
71. Du Pont annual report 1989, IDB annual report and accounts 1989-90 (pp 81) and extrapolation of figures.
72. *Irish News*, 16 April 1991.
73. During the early months of 1991 Mary Harney, the minister for the protection of the environment in Dublin, began to research the feasibility of a toxic waste inventory similar to the system employed in the US by the EPA.
74. "Toxic Waste Incinerator – A poisonous fiasco", *Derry Journal*, 23 April 1991.
75. See *Derry Journal*, 26 April 1991 and *The Sentinel*, 2 May 1991.
76. The full draft is on file in the DDEC.
77. MacKenzie, P. and Vaughan, R: The proposed national toxic waste incinerator for Derry; Briefing paper. April 1991.
78. "Derry Says No", *Derry Journal*, 3 May 1991.
79. "Minister treating Donegal with Contempt," *Derry Journal*, 3 May 1991.
80. Vaughan, Rosemary. Correspondence with Robert Allen 1991.
81. MacKenzie, P. See also *Derry Journal*, 18 June 1991.
82. McKenzie, P. Interview with Robert Allen, 1991.

83.   Wilie, R. Interview with Robert Allen, 1991.
84.   Ibid.
85.   Ibid.
86.   My thanks go to the press officers and civil servants in the IDB, the DoE (in Belfast) and the DoE (in Dublin) for this and earlier information.
87.   McKenzie, P. Interview with Robert Allen, 1991.
88.   Wilie, R. Interview with Robert Allen, 1991.
89.   As explained to Peter McKenzie by Richard Wilie.
90.   Wilie, R. Interview with Robert Allen, 1991.
91.   Harney, Mary. Interview with Robert Allen, 1991.
92.   Wilie, R. Interview with Robert Allen, 1991.

Chapter 2

INVERNESS AND NONTOX
A Tale of toxicity

Neil and Fiona Sinclair moved to the Nairn area in the summer of 1988. Neil had got a job in the local secondary school. Fiona's contract had ended on a Manpower Services Commission community programme. The clean Highland air and the country tranquility contrasted dramatically with their previous home in a town centre flat in Stirling. In January 1989 they bought their own house in Culloden, a new housing area five miles east of Inverness and a couple of miles north from the site of the infamous battle. They settled into their new home and Fiona managed to get a job at the Culloden Visitor Centre. While there she saw a small notice in the local paper, which mentioned that the then Secretary of State for Scotland, Malcolm Rifkind, had ordered a public inquiry into Highland Regional Council's refusal of planning consent for a "controversial" incinerator at the Dalcross industrial estate plant of Nontox.[1]

For me, the key words were "controversial", Nontox and incinerator – as it was a private company, it could only be burning industrial nasties of one description or another, and, because I'm a cynic, I thought the name must mean the complete opposite. There was an advert in the same edition about a meeting being held by Greenpeace on turning the Moray Firth into a reserve for dolphins. I was quite interested in this, but I was more interested in going along to see if I could find any "environmentally aware" person who would be able to tell me something about Nontox .

The meeting hall, in Millburn Academy, was packed, and I just stood up, when I got the chance, and asked if anybody knew about Nontox, because I reckoned that this would be a far bigger pollution hazard than the Regional Council's plans for piping raw sewage into the firth. A guy sitting along from me said he knew a bit about them, and I got talking to him, and another three people after the end of the meeting. One of those three was Sonia Jacks, who lives in Ardersier, the nearest village to the Nontox site, and who has since done her bit to help the campaign. The guy was Julian Clokie, who runs the Sea Vegetable Company over in Portmahomack, and he offered to take me round to see one of the tenants on the Industrial Estate.[2]

In June 1989 Fiona was introduced to the workers at Tommy Ross's scrapyard and was told about others who had been affected by Nontox. She paid other visits to the area over the summer, speaking to residents and to workers and tenants on the Dalcross Industrial Estate, in which Nontox is situated. "I felt a bit peculiar, going around asking these people to tell me about their experiences of this company. I felt a bit awkward, because, well, why was I doing it? I wasn't a journalist, or one of these professional environmentalists. Running campaigns isn't my thing, it still isn't."

A major problem for Fiona Sinclair was sorting out folklore from fact, as there was a wealth of stories about the company, very few of which would actually stand up at a public enquiry, either because they were hearsay, or because those who told them were unwilling to repeat them to a wider audience. This reticence was caused by a mixture of reasons – belief that the information could not be corroborated, and thus proven, belief that the authorities would not do anything anyway, and belief that they would be the victims of a backlash from either their employers or, in the case of the tenants on the industrial estate, their landlord, Highland Regional Council. The latter two reasons were rooted in the frustration which people felt at the local authorities' handling of the situation. Inverness District Council, who have the responsibility for monitoring Nontox, came in for the bulk of the criticism. John Horsfall, the owner of an aircraft maintenance firm, situated at the back of the Nontox plant, described their attitude:

We started phoning the Environmental Health. We noticed that whenever we did, the minute we did, in five minutes that thing was switched off. Four hours later two chaps would turn up and say "We hear you're having a problem with smoke – we can smell nothing." I said, "Well, why don't you bring some measuring equipment?" and the chap says he's got nothing, and I say I've got instruments costing £500 and I'm one man running a one man show, but those in the council with a great big budget hadn't a thing in order to monitor it, and they always turned up hours after you complained. We were shocked at the way we were treated – there was no concern.[3]

All these complaints were confirmed by other residents, tenants and workers on the industrial estate, and the specific accusation about the switching off of the incinerator was reiterated under oath by John (a committed Christian and lay preacher) at the subsequent public enquiry, yet no statement was issued by the District Council then or at a later date.

Despite the large number of complaints made (approximately 100 to the district council alone, by John Horsfall's reckoning, 19 of which were made by John himself), none of the statutory authorities showed much concern. The Civil Aviation Authority, HM Industrial Pollution Inspectorate, (HMIPI), the Health and Safety Executive and the district council were all contacted, but the local people did not feel they were getting anywhere. They had complained of nausea and lethargy, blinding headaches, runny eyes and sore throats, most of which they alleged were the result of smoke and fumes from the incinerator, but some were caused by fugitive emissions from off-loading of tankers.

Over the following months Fiona Sinclair attempted to get more information from the local authorities about the Nontox operation. She made an appointment to see John Greaves, Highland Regional Council's Deputy Director of Planning; Sonia Jacks accompanied her. The meeting itself raised more questions than it answered, but she was given a copy of the department's report to the Planning Committee, which recommended refusal of planning permission for Nontox's incinerator.[4] She later asked John Greaves for a copy of the Thomson report, and was told that, as this report was commissioned by Inverness District Council to investigate Nontox's activities, they were the proper authority to ask for a copy.

Fiona Sinclair's attempts to elicit the Thomson Laboratories Report from Inverness District Council's Director of Cleansing, William Fraser, reached farcical proportions. No, with no reason given, was the first response; the second gave an incorrect reference to an Act of Parliament[5]; the third response was written by the Director of Legal Services, D.R. Somerville, and arrived on the first day of the Public Inquiry.[6] The letter stated: "However, I can now see that the particular report concerned has been lodged as a production in connection with a Public Inquiry and therefore can now be treated as a public document."

Two further detailed letters were sent to the Directors of Cleansing and Legal Services after the enquiry was over, and receipt was made of "the usual two sentence reply" from the Director of Cleansing, and the final retort from the Director of Legal Services:

As you are aware, the need to maintain confidence in the report passed and therefore it was made available as a production by the District Council to the Regional Council for the Planning Appeal. I can well understand that you do not fully appreciate the planning process but that Report was always available for inspection through the Planning Department when it was listed as one of the Productions and as it was a Regional production would have been made available to you had you requested same. Public Intimation of this would have been given by the Scottish Office Enquiry Reporters Unit.[7]

"In my mind," said Fiona Sinclair, "this just amounted to deliberate obstruction on both their parts, even if it were the case that the report was available from the Planning Department. Of course I don't "appreciate" their planning processes, when, as public servants, it was incumbent upon them to inform me where I could get hold of this report. I could have been given this information in reply to my first letter, and I can only conclude that they had a reason for wanting to prevent members of the public from seeing the report, prior to the enquiry. It so happened that John Martin, of the Highland Green Party, gave me a loan of a copy he had elicited from the HRC Planning Department, but I only had this over the weekend."[8]

Workers, tenants and residents were all biding their time, waiting for Highland Region to push their case for them at the enquiry and they put their faith in the enquiry system. A new development came seven days after Nontox was finally arm-twisted by the region into applying for retrospective planning permission for the incinerator, when Lanstar, a Manchester chemicals recovery firm, bought Nontox on 8 August, 1988. The region were sufficiently worried about this to ask objectors to the planning application not to hold back on complaints, as they rightly feared would happen, once the public knew that the Inquiry was pending. "Prior to the enquiry the Regional Council said, 'see if you can find some cause to complain because it's been so quiet. When it comes to the Inquiry Nontox can stand up and say there have been no complaints about them for the last six or 12 months'," George Cocker noted.[9]

The Public Inquiry began on 24 October 1989, and was held in the Regional Buildings in Inverness. Highland Regional Council's Principal Solicitor, Alan Simpson, presented the case for the region, with planning and technical support being provided by John Greaves and Dr Peter Kayes, of Environmental Dynamics. James MacKay, the Director of Environmental Health; John Williams, the Deputy Director of Cleansing; and Donald Somerville, the Director of Legal Services, were also present through most of the Inquiry. Alexander Phillip presented the case for Lanstar, with technical support by Cremer and Warner, the chemical engineering firm. Dr John Parker, the Technical Director of Lanstar, and Bill Lockhart, the Operations Manager, were also present throughout the inquiry.

Until the Inquiry began the community did not know what kind of materials had been burnt in the Nontox incinerator. The Thomson Report alluded to the possibilities:

> Adequate records for the site operations were not available. Since there is no site weighbridge, confirmation of wastes received on site, held in transit and disposed of by incineration could not be provided.
>
> There are no records available of any analyses of the compositions of the wastes accepted on site and/or consumed in the incinerator. Only records of "special" waste are kept in an accessible form.
>
> The descriptions of the wastes received are generally non-specific and too vague for pollution control purposes.

The consultants recommended that "all wastes submitted to Nontox should be subjected to independent analysis to check for other species, particularly trace metals, bound nitrogen, PCB's, semi-conductor dopants, chlorinated and fluorinated hydrocarbons, iodine and bromine. It would also be sensible for the operators to determine the calorific value of the waste and the water content."

The district council had assumed that the wastes being delivered under the consignment note system to Nontox were, in line with the normal use of this system, "special" wastes.[10] In the first 18 months of the incinerator's operation, only 12.5 per cent of wastes consigned to Nontox may have been reclaimable using their oil recovery process.[11] Although records "of the types and quantities of wastes delivered to and the residues removed from the site" were supposed to be kept, as a condition of their waste disposal licence,[12] Nontox's own records were extremely poor. Of all the consignment notes sent to IDC's Cleansing Department, 27 per cent did not contain quantitative information of the composition of the waste and 32 per cent contained errors.[13] This placed Nontox's claim that they kept the waste disposal authority informed of their activities through this system in a rather dubious light. It was, in fact, stated by a Lanstar witness at the enquiry that "consignment notes are used when people don't know how to categorise waste".[14] Although the waste disposal licence only allowed the incineration of a range of hydrocarbons which were byproducts of their existing oil recovery process, paint wastes commonly featured on consignment notes,[15] in breach of licence conditions.[16] Lanstar continued to accept this paint waste, until after the publication of the Thomson Report.[17]

Paint wastes and sludges featured prominently in Nontox's consignment notes. In November 1987, Cremer and Warner recommended strongly that Nontox should cease the practice of incineration of sludges and non-pumpable solids. However, it can be seen from the analysis of the consignment notes that Nontox and latterly Lanstar Waste Treatment have continued to receive and incinerate waste materials containing significant quantities of sludge. Up until the end of the first licensing period (end of August 1987) a substantial proportion of the wastes incinerated at Dalcross were classed as containing paint wastes and sludges. This pattern was repeated again in 1988. The review of those consignment notes which were issued during the last year of operation of the incinerator reveals that the quantity of paint waste notified by the main consignor, Lanstar Wimpey Waste, had fallen to zero, whereas over the same period, notifications by the same company for sludges increased correspondingly such that over 80 per cent of all samples received at Nontox were confirmed by the consignment notes to be contaminated with sludges.[18]

Although Lanstar's QC made a bold attempt in his summing up to dispel the poor image of Nontox's operations, which he claimed was "a matter of historical fact only"[19], Dr Peter Kayes, Highland Region's technical adviser at the enquiry, pointed to the fact that Nontox's latest planning application for their incinerator included the capability of burning non-pumpable semi-solid wastes.[20] To add further emphasis to the link between paint wastes and sludges he had earlier alluded to, he explained:

Much of the waste material disposed of in the incinerator in the past year has been classed by the consignor as mixed solvents. Typically, these appear to have contained the following components as the main solvents:

i)   Isopropyl Alcohol
ii)   Isopropyl Acetate
iii)   Xylene
iv)   Toluene

These solvents find major industrial uses and are ubiquitous in the paint industry.

Dr Kayes believed that this accounted for the preponderance of these solvents in the waste streams.[21]

Nontox can be traced back to its inception at a farm steading at Delnies, near Nairn, and the planning consent given in 1980 to its then owner, Mike Jacobs, to relocate the company's oil recovery and storage operations to a site at Dalcross Industrial Estate. Nontox was issued two waste disposal licences for its incinerator by the waste disposal authority, the Cleansing Department of Inverness District Council, over a four year period, despite the fact that planning permission had not been obtained for the incinerator. Nontox was first granted a waste disposal licence in 1985 for the incineration of residues from the existing oil recycling processes operated at the Dalcross site.[22]

According to the Director of Cleansing of Inverness District Council, he "was informed by the Planning Department [of Highland Regional Council] that, from the information given to them by Nontox, planning consent was not required for the installation of the incinerator".[23] The Director's informant was Jim Falconer, who, in his observations on Nontox's waste disposal licence application for the incinerator of 30 May 1985, had written: "This was discussed with Nontox Ltd and from the information given they were advised that Planning Consent was not required." He had signed this brief comment on behalf of the Divisional Planning Officer. However, according to the Planning Committee report of 26 October 1988: "Late in 1985, Nontox made enquiries to Inverness District Council seeking modification to licensing conditions to accommodate incineration. This was the subject of consultation with the Planning Department who responded to the effect that the original planning consent was narrow in scope and did not include incineration".[24] This would seem to be in response to an application which Nontox made to the waste disposal authority to extend the range of wastes to be burnt.

On that occasion, the WDA [Waste Disposal Authority] consulted with HRC's Director of Law and Administration who in turn contacted the Department of Water and Sewerage and the Headquarters office of the Planning Department rather than the Inverness Divisional office. The Planning Department responded to the WDA by letter of 18 November 1985 and also wrote in similar terms direct to Nontox, advising of the need for planning permission, as your new proposals represent a significant departure from the activities previously approved. Copies of these letters were passed to the Inverness Divisional Planning office. Nontox subsequently withdrew its application for the proposed modification to its recently granted licence."[25]

Such an obvious contradiction arising between local government departments seemed incredible – how could such a "misunderstanding" arise? But from this misunderstanding, more were to follow. The first waste disposal licence ended, two years later, on 30 September 1987. The incinerator had been installed in February and commissioned in March 1987. When Nontox applied for its second waste disposal licence in 1987, the Cleansing Department again issued a waste disposal licence for a period of two years (ending September 1989), apparently "under the impression" that Nontox would lodge a fresh planning application.[26] The communities of Dalcross and Ardersier, the nearest villages to the Nontox site, were ignorant of the details of HRC's dealings with Nontox. Some were plainly shocked that Nontox never had planning permission for their incinerator. According to Highland Region's Planning Department, Nontox "have been informed of the need for planning permission on repeated occasions both before and after the operation of the incinerator began".[27]

Falconer (or rather Highland Region's Inverness Divisional Planning Office, which is located in the district council's offices) continued to complicate the issue. On 21 March 1986, he wrote a letter to the Managing Director of Nontox:

> OIL INCINERATION PLANT – NONTOX LTD,
> DALCROSS INDUSTRIAL ESTATE
> I refer to your letter of 20th March regarding the above and would confirm that as the plant is to be contained wholly within an existing building, a formal application for planning consent will not be necessary for this installation.
> It should be noted that this plant should not be used for any purpose which would contravene the conditions of your planning consent granted on 2nd June 1980.

Common sense alone would say that the unequivocal nature of the first paragraph could override the conditions laid out in the 1980 planning consent which, although not making mention of incineration, does not specifically exclude it. However, planning consent is deliberately vague, and material changes must have their own planning permission. Much of Nontox's case, as summed up by their Q.C. at the Public Inquiry, rested on the question of whether or not the incinerator did in fact constitute a material change in the original planning permission. This was combined with what the Q.C. saw as inconsistency on the part of the Planning Department in that Mr Falconer was "doing his job" when he wrote this letter, was "accustomed to doing" this, but had "no authority" to do so. Furthermore, Nontox could not think there was any "irregularity" about this letter which he considered written "in clear terms" and which he claimed was not denied by HRC.[28] The situation was further complicated when, in January 1987, Nontox applied to HRC for planning permission for a hot air discharge flue, which was described as being "from industrial waste burning furnace" and for the erection of storage tanks for oil contaminated liquids. Both applications were approved at Divisional Planning level in April of that year, and were subsequently implemented.[29]

In June 1987, Nontox submitted an application for planning permission for the incinerator. Pressure was applied by HRC after Nontox had submitted a new application for a Waste Disposal Licence in which the range, nature and difficulty of the wastes it was proposed to handle would increase significantly. It was pointed out that, as a result, the incinerator would require not only planning permission, but separate licensing by HMIPI.[30] Mr Mike Jacobs, Managing Director of Nontox at that time, withdrew this application in December 1987[31], as he did in March of 1988 with the second application for a waste disposal licence.[32] "Subsequent meetings, discussions and correspondence led HRC to believe that a fully detailed amending application and a Consultants' Assessment were in course of preparation."[33]

Nothing else happened until 1 August 1988 when Nontox submitted a new application for retrospective planning permission for its incinerator, in response to pressure from HRC, which threatened formal action.[34] The planning application was refused in November 1988, when HRC also took out an enforcement notice against Nontox.[35] Nontox, however, appealed against these decisions to the Secretary of State for Scotland and were able to continue operating their incinerator until September 1989 when their waste disposal licence expired – the District Council had finally decided that they could not, in the circumstances, provide another waste disposal licence.

Reasons stated at the Inquiry by Dr John Parker, a director of Lanstar, for the takeover of Nontox were because the Highlands represented "an interesting little market" without much competition.[36] After Lanstar's takeover, the nuisance of fumes from the incinerator diminished. Local people surmised at the Inquiry that this had less to do with a desire to improve the local environment than with an attempt to win their appeal. Lanstar successfully steered the Inquiry's attention away from Nontox's history of public complaints prior to the takeover. However, Highland Regional Council did manage to produce some evidence to back up their own claim that, whilst improvements had been made, operational standards were still wholly unsatisfactory.

Lanstar's relationship to Nontox prior to its takeover of the Dalcross firm is, from the available information, an interesting one. Within three or four months of the commissioning of the incinerator, consignments were being received from Lanstar Incineration and Lanstar Wimpey Waste (the latter is a joint venture company between Lanstar and Wimpey Waste Management). "In the first year of operation, these shipments accounted for 71 per cent of the total quantity shipped in that year to Nontox from all sources. This fell to about 60 per cent in 1987/88 and rose again to 91.2 per cent in the last year of operation after Nontox had been taken into the ownership of the Lanstar Group of Companies."[37]

At the Inquiry Lanstar also revealed that it had undertaken testing of some samples of loads for Nontox, as Nontox did not have any kind of laboratory or indeed a chemist to carry out such testing.[38] The circumstance which excited most interest at the Public Inquiry was the revelation that Lanstar, 14 months prior to its takeover of Nontox, had been gifting support[1] fuel to Nontox. It had not only paid Nontox to take the support fuel, but had paid for its transport as well.[39] Evidence was given by a waste disposal expert, on behalf of Lanstar, that such a practice was normal in the industry.[40] HRC's principal solicitor, Mr Alan Simpson, did his best to point out that whether or not such substances were used as support fuels, it did not alter the fact that this had involved a breach of the company's waste disposal licence, because the substances being used as support fuels were solvents.[41] Lanstar did not satisfactorily answer why, given that "gifting" of support fuel is a normal feature of the waste disposal industry, it did not gift this fuel to other waste disposal companies nearer to their base. It was necessary that this support fuel was free, to enable the commercial recovery of oily wastes. Lanstar contended that the incinerator was installed because of the restrictions imposed on Nontox by the Cleansing Department on the amount of liquid wastes to be disposed of at the Council's tip.[42] This illusion was shattered by the evidence that in 1987 the total quantity of liquid wastes disposed of at the tip amounted to 44,000 gallons, which in the first six months of 1988 rose to 164,000 gallons.[43]

Dr Peter Kayes, an environmental consultant commissioned by HRC to provide scientific evidence on their behalf at the Inquiry, alluded to "the apparent resistance of the appellants to independent analysis

---

[1] Support fuels are combustible wastes sometimes used to maintain heat in the incinerator.

of waste consignments prior to incineration".[44] He also revealed his recent discovery that seven copies of certificates of analysis of waste consignments provided by Lanstar were based on samples taken over a relatively short period of time – specifically from August 1988 until January 1989, with the exception of one sample taken in August 1989, and therefore did not represent a random selection of waste consignments.[45] Although Lanstar denied that any of the wastes being dealt with at Dalcross were "special"[46], HRC found that a few consignment notes confirmed their presence in the wastes being handled.[47] The calorific value of wastes, an important piece of information for their successful incineration, had not been provided by Lanstar on the consignment notes. This is not surprising if Dr Kayes' theory as to the true nature of waste disposal at Nontox is considered. "In order to consume increased quantities of High Calorific Value Special Wastes, larger and larger volumes of watery wastes are required to be injected into the incinerator to provide a suitable heat sink and cooling load to protect the fabric of the combustor and its refractory lining".[48] Several locals noted that the flames from the Nontox stack flared sometimes as high as 20 feet – a symptom of the use of too much High Calorific Value material in an incinerator.

In their defence of Nontox, Lanstar cited the absence of any objection by HMIPI or the Health and Safety Executive to the granting of planning permission.[49] The company also tried to shift the blame for complaints onto two other local polluters, Highland Forest Products and Hambleside Plastics. Local people were, however, adamant that Nontox were the offenders.[50] On the final day of the Inquiry, Lanstar revealed its future plans for the operation of the Nontox plant. The company's Q.C. started to make a case for renewed licence conditions which included an extension of the wastes covered by their waste disposal licence. He added that no tests would be done for dioxins, as this was considered unnecessary and too expensive.[51]

Lanstar's own expert witness, Mr Wiltshire of the consultants Cremer and Warner, stated that none of the loads being delivered to Nontox would require HazChem codes to be displayed on the lorries,[52] yet locals frequently saw vehicles entering the Nontox site displaying HazChem signs.[53] The workers at the BP airport fuel dump, whose office is directly at the back of the Nontox site, seem to have come in for the worst effects from Nontox's activities. One entry in the log book compiled by the workers is of particular note:

After getting fallout from Nontox yesterday I had to get the engineers to cover for me while I went home to get a shower as the itch had started to turn to a burning sensation. When I went into the shower my skin felt like it was on fire and I could hardly touch it with the towel as it was very tender. After a while it cooled down but later at night about 20.30 approx. I was out at a function and my face started to burn again and come out in little pinhead blisters which made my skin damp. At night I could hardly sleep so I put on some baby lotion and that helped. Today my face feels alright but [the] skin is tight feeling. What lasting affect could this have on my health now or in the future?[54]

There were only two HazChem codes noted on the lorries entering the Nontox site in the BP workers' logbook. Fiona Sinclair wrote to the Toxics division of Greenpeace UK prior to the Inquiry to obtain information about these codes, and she included this information in her submission to the Inquiry, which was unchallenged by Lanstar's QC. On 17 February and 2 March 1988, tankers were noted discharging at Nontox, carrying the code "3 WE 7010". The code is broken down to reveal: "3 WE is the firefighting code, meaning a fire should be fought with foam. Protective clothing should be worn, and the spill should be contained. There is a danger of violent reaction or explosion. 7010 means hazardous liquid flammable waste, with a flashpoint of less than 21°C (which probably means industrial solvents)."

The other HazChem code was "2X 7017", noted offloading on 11 February and 7 April 1988, and broken down as: "2X is a firefighting code, meaning a fire should be fought with water fog, rather than water jets; protective clothing should be worn and the spill should be contained, not allowed to leak into drains. 7017 indicates hazardous toxic liquid waste."[55]

HRC expressed concern, in relation to the second code, that the Nontox site would allow substances like solvents to drain very quickly down through the soil into watercourses in the event of a spill.[56] Mr Head, the HMIPI District Inspector who acted as technical adviser to the Reporter throughout the Inquiry, should have known that one definition of a "special" waste is a waste having a flashpoint of 21°C or less. The evidence submitted on the first HazChem code would have elicited some concern, and consequently some attention in the Reporter's letter of 26 April 1990, which noted the Reporter's inclination to give planning permission to Lanstar for their incinerator at Dalcross.

Some exasperation was obviously felt by HRC's Principal Solicitor, Alan Simpson, at the Inquiry, particularly at what he fittingly described as the "moving target"[57] of Lanstar's plans. Dr John Parker stated, for instance, that the minimum flashpoint of the "support" fuel would be 32°C; in Mr Philips' summing up, this was given as 25°C, and in Lanstar's later written Comments on the Proposed Licence Conditions, it was given as 21°C. The feedrate for the incinerator was also in a state of flux – Nontox had originally applied for a maximum feedrate of 5 tonnes per hour[58], but were informed by IDC that they would only be allowed to operate to the maximum set by the incinerator manufacturers, which is 0.55 tonnes per hour.[59] Despite this, their 1988 planning application confirmed that the maximum design throughput would be 3 tonnes per hour[60], which is in excess of the 2.268 tonnes per hour feedrate recommended by Dr Kayes for watery, low calorific value wastes.[61]

The company had no plans to fit filters or air scrubbing equipment to its incinerator stack which meant that the acid gases, fume and particulate matter produced would be discharged direct to the atmosphere.[62] Instead of using pollution abatement equipment for the incinerator, Lanstar decided that, as "special" wastes would not be incinerated[63], and as they would not describe the support fuels as hazardous, pre-acceptance screening of wastes was deemed sufficient to ensure satisfactory operation of the plant.[64]

It was HRC's contention that the past performance of Nontox before and after Lanstar's takeover did not inspire confidence in any future operations. Dr Kayes concluded that the Nontox laboratory and the company staffing levels were not adequate for the analysis of oils, never mind solvents.[65] Nontox had not furnished Kayes with the results of any stack gas analyses[66], and, according to Mr Wiltshire, there was no emissions data, because no emissions were observed or monitored.[67] In addition to this, there were "37 tons of material delivered by BP and which have not been accounted for in the statistics on a volume basis."[68]

Quite apart from any other wastes which may have produced dioxin emissions when incinerated, the Thomson Report recorded that 12,217 gallons of semi-conductor wastes (i.e. 2.4 per cent of the total waste consigned) had been received on site by Nontox.[69] And apart from the relatively minor transient effects reported by local people, which they attributed to Nontox's operations (mostly the incinerator), there were reports of effects on animals. Three deformed lambs were born on Woodend Farm, from ewes grazing in a field opposite the Nontox plant. John MacLennan, the farmer, claimed that this was the first time he had experienced this, in over 24 years of farming this land. A deformed calf was also born on Baddock Farm.[70] A pony, apparently in the best of health, which grazed on a field beside the Nontox site, was found dead. The pony's owner, Tommy MacDonald, is Nontox 's nearest residential neighbour. MacDonald recounted: "I breed birds [budgies, canaries and cockateels], and I had no success with breeding whatsoever. Since that infernal thing started the chicks were dead in their shells – that was it, nothing. . . . About three weeks after it stopped I had a nest of budgies, which seems rather a coincidence to me."[71]

These events worried local campaigners. They had not been reassured by the measures which the waste disposal authority had taken. The WDA's soil sampling around the Nontox site, conducted over a year after the closure of the incinerator, was criticized.[72] Scottish Community Organisations Opposed to Toxics (SCOOT), the local campaign group set up to fight the recommencement of incineration at Nontox, said, "only three samples were taken at distances of less than three hundred metres from the site and only the turf on the surface was tested. As Inverness District Council waited almost a year after the incinerator stopped working before taking samples for analysis, we can hardly expect the pollutants to have remained in the top layer of sandy soil".[73]

For campaigners like Fiona Sinclair, the regulatory authorities' attempts to deal with Nontox and to explain the problems in and around the Dalcross area amounted to little more than "a sop to the community council and Highland Green party".[74] Nobody, it appeared, took the problem seriously.

I wouldn't say they didn't take it seriously. The local councillors were entirely guided by advice from their officials and were not prepared to look beyond government guidelines to the more basic problems we had alluded to as a group about toxic development such as the Nontox incinerator.

We wrote to the Department of Agriculture and Fisheries for Scotland and they provided a totally misleading response to our questions concerning testing of local milk for dioxins. They claimed the composition of Nontox's fuels were specified by and regularly checked by Inverness District Council. We know this to be wholly incorrect. They must have thought we were buttoned up the back by claiming that the chlorine content of the wastes was for the most part below 0.1 per cent. It is a fact that only one out of the seven samples given to the District Council had such a chlorine content. They further claimed that PCBs were not incinerated at any time whereas we know that Lanstar have had the stupidity to list PCBs under the metals category, in their certificates of analysis.

The Highland and Islands Development Board weren't much better. They claimed that funding of Nontox had ceased in 1985 but it appeared that it hadn't as Falconer's letter of March 1986 was apparently used by Jacobs to solicit funding from the HIDB.[75]

Lanstar even managed to have a fire while the incinerator has been out of operation, at the end of July of last year. At the time we got at least two versions of how it had started – a local paper said that a drum of waste oil had ignited, and we heard from another source that a tank lining had gone on fire. Nobody in the local media took it upon themselves to clarify the situation, indeed the only media reports were on the local commercial radio station, and a six line column in one of the freebies, and if I hadn't got word to them that there were fire engines from Inverness out at it, there would have been nothing at all.[76]

Sinclair was highly critical of the standards of the local news coverage, which she claimed lacked basic accuracy and fairness in its handling of the Lanstar issue.

We haven't got anything wrong so far, yet any information we care to present is framed as an allegation or claim, and they will always quote the council and the company, as authoritative sources, giving them the last word and guaranteed access to print. Three local papers published a Lanstar public relations release virtually word for word, without coming to us for any comment, yet if we put out a press release, the media will make damn sure that they give a "balanced" view of things.

Sinclair admitted that the lack of active support from the local community had played into the company's hands, and it was, in some respects, no wonder that there was a lack of interest amongst the local media and politicians.

But then again, the Clydeside [see Chapter 6] campaigns show that the whole media thing is a self-fulfilling prophecy. The active campaigning support of the *Evening Times* has made their campaigns. And then again, Lanstar, in one manifestation or another, has been around for a number of years – the local community successfully resisted plans for Nontox to use part of the Ardersier Carse as a toxic dump some years back – so I feel that battle fatigue has definitely set in.

Sinclair pointed to the recent legislative, regulatory and policy changes which have taken place as evidence of a strong push by the government to locate toxic development in Scotland, particularly in the rural areas. Indeed, the timescale of various events surrounding the Nontox story adds an interesting dimension to what might otherwise seem to be a local problem. Apart from the consolidation of the Manchester connection, which took place within a week of Nontox making its final, and ultimately successful, application for planning permission, there is a string of events which Sinclair insists is no mere coincidence.

The Nontox Public Inquiry ended on 3 November 1989. Five days later came the announcement that the Scottish Office proposed changes in planning appeals procedures which "would allow legal precedent to be taken into account when an appeal was being considered in spite of the tradition [enshrined in Scots law] that each case should be considered on its own merits". Furthermore, "The Scottish Office document lists examples of unreasonable behaviour which might give rise to the award of expenses in a planning case. These include failure by a local authority to take account of the statements of Government policy in departmental circulars or any precedents of which they should have been aware". It was also reported that the Scottish Office proposed to introduce fees for planning appeals.[78] At the beginning of January 1990, it transpired that the legislation to set up Scottish Enterprise, and Highlands and Islands Enterprise (to replace the previous development agencies of the Scottish Development Agency and the Highlands and Islands Development Board) would allow the Secretary of State for Scotland to grant planning permission to any development with the backing of HIE, without being obliged to take any account of what HRC said.[79] This particular part of the (well advanced) legislation was shelved, after vociferous protest.

The Reporter of the Nontox Public Inquiry, John H. Henderson, published his letter on 26th April 1990, indicating his inclination to give planning permission to Lanstar for the incinerator, subject to agreement on planning conditions being reached between the company and HRC and IDC. In the final paragraphs of his letter, Henderson stated: "it cannot be right that a development proposal should not be allowed to proceed because it is alleged that the controlling agencies are unable to adequately meet these obligations."[80] He went on to affirm his confidence in the controlling agencies' ability to monitor the incinerator, with "the likelihood that people living and working nearby will not be slow to draw attention to things going obviously wrong, I do not believe it realistic to assume that any unsatisfactory operation would be of more than a very short duration before the problem would be corrected."[81] Local people alleged that, at the pre-Inquiry meeting, Henderson claimed that he was not an authority on toxic waste disposal. They were angered to discover later that he had been the Reporter on the Public Inquiry for the Brackmount Quarry dump in Fife, and that the company applying for planning permission for this dump, Wimpey Waste Management, had a commercial connection to Lanstar, through their joint venture company, Lanstar Wimpey Waste.[82] Two months after the publication of the Reporter's letter, Nontox was renamed Lanstar (Scotland)[83], prompting comparisons with Calderhall/Windscale/Sellafield.

On 7 November 1990, the Secretary of State for Scotland announced his approval of Highland Region's Structure Plan Review, the authority's major development policy document, with a number of "modifications".[84] At the time, those policy modifications which received most media interest were the deletion of the region's policies against the establishment of a national nuclear waste repository in its area, its presumption against the granting of planning permission for geological exploration for potential waste disposal sites and its plans for expansion of skiing facilities into Lurcher's Gully in the Cairngorms. One other modification to the Structure Plan Review which received very little press coverage was the deletion of the region's policy of resistance to the importation of toxic wastes into its area, which Highland Region had used as one of the three main factors in its case against the Nontox incinerator.[85] One other significant change went completely unnoticed by the authorities and the media. Sinclair did, however, notice the Secretary of State's deletion of all environmental safeguards from coastal quarries served by sea or rail only.[86] There are a number of coastal "super quarries" which are presently proposed for Highland Region, the Western Isles, Orkney and Shetland, and she believes that if the Scottish Secretary is unconcerned about such safeguards existing for the mainland, the same will surely apply, either now or in the near future, for the isles. The knowledge which puts this particular change into perspective is a Friends of the Earth (Scotland) report into landfill sites in Scotland, which stated that Glensanda Quarry, on the north shore of Loch Linnhe in Lochaber District, already contains "difficult" wastes, even though extraction of minerals is still taking place.[87] Highland Region seem unconcerned about the possible environmental effects of their own minerals policy, which encourages the exploitation of barytes, copper, diatomite, gold, lead, molybdenum, potash, zinc and uranium.[88] On 14 November 1990, seven days after Malcom Rifkind's announcement concerning HRC's Structure Plan Review, it was reported that the Lanstar group had been bought by Charterhouse Venture Funds[89], who had previously owned ReChem.[90]

At the end of 1990, Sinclair was informed by Cathy Russell of CACI of a change to a Statutory Instrument called the Heavy Industrial Use Classes Order, which was notified, at that time, in a COSLA (Convention of Scottish Local Authorities) document. This proposed change has advanced further down the road towards implementation, and will mean that an incinerator can be erected by any company which has a site classed as having a heavy industrial use, without obtaining planning permission from the local authority. Only HMIPI's permission would be needed.[91] Added to this centralization of planning control,

Sinclair was informed by a local journalist of a regulatory change which will mean that monitoring of some kinds of incinerator will be taken over entirely by HMIPI, whose sole base is in Edinburgh. The Environmental Protection (Prescribed Processes and Substances) Regulations 1991 are due to be implemented on 1 April 1992. The change will also mean that companies such as Lanstar will *not* be required to have a waste disposal licence issued by the local district council.[92] With such favourable conditions being created by the government for the waste disposal industry, it is hardly surprising that Lanstar announced the sale of its chemicals division at the end of June 1991, to enable them to concentrate on their core business of waste treatment.[93]

Unlike other campaigns, SCOOT have pursued an anti-incineration line, their reasoning being that any other stance would have been immoral, and, in the end, ineffectual. SCOOT has made a point of making direct contact with other groups, and has given and received information. For Sinclair, "NIMBYism is self-defeating. If incineration of toxics isn't okay for your community, then don't try and fob it off on someone else, and don't expect sympathy when this approach fails, as it is bound to do."[94] For this reason, the limbo-like situation which prevents HRC from appealing, in the absence of a final decision being made, has not really affected SCOOT's campaigning, although it does give them cause for some concern. The situation obviously causes concern to HRC's Principal Solicitor, Alan Simpson, who has not so far received a clear statement as to the precise nature of the changes envisaged by Lanstar as a result of the proposals submitted by their consultants, Cremer and Warner, at the Public Enquiry. HRC therefore finds that it is "not possible to finalise draft planning conditions". The additional danger of further control being exerted by the Scottish Office is fully appreciated by Simpson, who is "conscious of the possibility of the Reporter taking this matter out of the Council's hands, and that is precisely why I have been at pains to ensure that the ball is placed firmly back in the developers' court".[95] In mid-November 1991 Lanstar decided to decommission the incinerator, but Fiona Sinclair said that this did not mean that the campaign had ended.

Meantime, SCOOT know they cannot rely on others to champion their cause, or to take the campaigning initiative. In Sinclair's words, "A major part of my whole philosophy of life is summed up in a quotation attributed to Burke – 'For evil to succeed, it is only necessary for good men to do nothing'. Leaving aside the gender bias, it says it all."[96]

Notes and references

1.   *Inverness Courier*, 2 June 1989.
2.   Sinclair, Fiona. Interview with Robert Allen, 1991.
3.   Horsfall, John. Interview with Robert Allen, 1990.
4.   Highland Regional Council Planning Committee report, IN/1988/812, 26 October 1988.
5.   Fraser, William. Letters to F. Sinclair, 28 September 1989 and 5 October 1989.
6.   Somerville, D.R.. Letter to F. Sinclair, 23 October 1989.
7.   Fraser, William. Letter to F. Sinclair, 14 December 1989 and Somerville, D.R. Letter to F. Sinclair, 20 December 1989.
8.   Sinclair, Fiona. Interview with Robert Allen, 1991.
9.   Cocker, George. Interview with Robert Allen, 1990.
10. *Inverness Courier*, 27 October 1989.
11. Highland Regional Council Planning Committee report, 26 October 1988, paragraph 5.2.
12. Inverness District Council Environmental Health Committee minutes, Appendix 1, condition 6, 12 August 1985.
13. Dr Peter Kayes' Precognition for the Public Inquiry on behalf of Highland Regional Council, paragraph 9.6.
14. Mr Wiltshire, Cremer and Warner consultant, inquiry testimony and as part of F. Sinclair's unchallenged Inquiry submission.
15. Dr Peter Kayes' Precognition for the Public Inquiry on behalf of Highland Regional Council, paragraph 5.11.
16. Ibid, paragraph 5.14. In the light of this one wonders why Inverness District Council did not take any legal action against the company (see IDC Environmental Health Committee meeting minutes, 11 October 1988, paragraph 1572 i) b) and Nontox/Lanstar Inquiry Reporter's Letter, 26 April 1990, paragraph 130), and why they did not think they had sufficient evidence to do so (letter from IDC Director of Administration to HRC Director of Planning, 18 October 1988).
17. F. Sinclair's unchallenged Inquiry submission.
18. Dr Peter Kayes' Precognition for the Public Inquiry on behalf of Highland Regional Council, paragraph 5.11.
19. *Inverness Courier*, 7 November 1989.
20. Dr Peter Kayes' Precognition for the Public Inquiry on behalf of Highland Regional Council, paragraph 7.3.
21. Ibid, paragraph 9.5.
22. Inverness District Council Environmental Health Committee minutes, 12 August 1985, Appendix 1.
23. Fraser, William. Letter to Sir Russell Johnston MP, 7 September 1988.
24. Highland Regional Council Planning Committee report, paragraph 2.2, 26 October 1988.
25. Nontox/Lanstar Inquiry Reporter's Letter, paragraph 11, 26 April

1990.

26. Nontox/LanstarInquiry Reporter's Letter, 26 April 1990, paragraphs 113 and 16 and taped telephone conversation between F. Sinclair and Cllr J. Cattell.

27. Summary Response to Grounds of Appeal (HRC Production), paragraph 2.26.

28. F. Sinclair's notes from Inquiry, see also *Inverness Courier*, 7 November 1989.

29. Nontox/Lanstar Inquiry Reporter's Letter, paragraph 13, 26 April 1990.

30. Dr Peter Kayes' Precognition for the Public Inquiry on behalf of Highland Regional Council, paragraph 5.2.

31. Nontox/Lanstar Inquiry Reporter's Letter, paragraph 17, 26 April 1990.

32. Dr Peter Kayes' Precognition for the Public Inquiry on behalf of Highland Regional Council, paragraph 5.14

33. Nontox/Lanstar Inquiry Reporter's Letter, paragraph 17, 26 April 1990.

34. Highland Regional Council Planning Committee report, paragraph 2.4, 26 October 1988.

35. Nontox/Lanstar Inquiry Reporter's Letter, paragraph 18, 26 April 1990.

36. *Inverness Courier*, 27 October 1989

37. Dr Peter Kayes' Precognition for the Public Inquiry on behalf of Highland Regional Council, paragraph 9.7

38. Nontox/Lanstar Inquiry Reporter's Letter, paragraph 19, 26 April 1990 (Lanstar's own lab does not have the capability to test to p.b.b., which is necessary for e.g. PCBs, but have such testing done "on an intermittent basis" elsewhere).

39. *Inverness Courier*, 27 October and 7 November 1989

40. Mr Wiltshire, Cremer and Warner consultant, Inquiry testimony and Nontox/Lanstar Inquiry Reporter's Letter, paragraph 76, 26 April 1990.

41. Dr Peter Kayes' Precognition for the Public Inquiry on behalf of Highland Regional Council, paragraph 5.5.

42. *Inverness Courier*, 27 October 1989.

43. Fraser, William. Letter to HRC Director of Planning, 4 October 1988.

44. HRC Statement of Observations Relative to Appeal..., paragraph 4.2, no.6.

45. Dr Kayes' Inquiry testimony and Statement by J.E. MacKay, Director of Environmental Health (samples taken on 8 and 18 August 1988 – two taken on latter day, 6 and 24 October 1988, 11 January and 10 August 1989).

46. *Inverness Courier*, 27 October 1989.

47. Dr Peter Kayes' Precognition for the Public Inquiry on behalf of Highland Regional Council, paragraph 9.3.

48. Ibid, paragraph 7.1.
49. Highland Regional Council Planning Committee report, 26 October 1988, paragraphs 4.5 and 4.6.
50. *Inverness Courier*, 31 October 1989 (it is interesting to note, from IDC Environmental Health Committee minutes, paragraph 2014, that consultancy positions were reversed when Highland Forest Products employed Thomson Laboratories as their consultants and Cremer and Warner were employed by IDC).
51. F. Sinclair's notes from Inquiry and Comments on Proposed Licence Conditions.
52. F. Sinclair's unchallenged submission to Inquiry (from a direct cross-question).
53. BP workers log (part of F Sinclair's Inquiry submission) and interjection by J Horsfall, reported in *Inverness Courier*, 31 October 1989.
54. BP workers' log
55. Gibbens, John. Letter to F. Sinclair, 11 October 1989.
56. Dr Peter Kayes' Precognition for the Public Inquiry on behalf of Highland Regional Council, paragraph 5.5.
57. *Inverness Courier*, 7 November 1989.
58. Dr Peter Kayes' Precognition for the Public Inquiry on behalf of Highland Regional Council, paragraph 4.3.
59. Ibid, paragraph 4.5.
60. Ibid, paragraph 7.5.
61. Ibid, paragraph 4.5.
62. Ibid, paragraph 8.2 and Comments on Proposed Licence Conditions.
63. Comments on Proposed Licence Conditions (3)f.
64. Comments on Proposed Licence Conditions (3) and *Inverness Courier*, 27 October 1989.
65. Dr Peter Kayes' Precognition for the Public Inquiry on behalf of Highland Regional Council, paragraph 12.6.
66. Ibid, paragraph 7.7.
67. F. Sinclair's unchallenged submission to Inquiry (from a direct cross-question of Mr Wiltshire).
68. Dr Peter Kayes' Precognition for the Public Inquiry on behalf of Highland Regional Council, paragraph 9.4.
69. Ibid, paragraph 9.2
70. MacLennan, John. Discussion with F Sinclair, Madeleine Cobbing of Greenpeace UK and Scott and Sam Medwid, June 1990 and Highland News 17 February 1990.
71. MacDonald, Tommy. Interview with Robert Allen, 1990.
72. MacKay, J.E. Letter and enclosures to Mrs Mary Cocker, 4 December 1990 (the SCOOT press release was issued in response to the press statement issued by IDC and the enclosures, which detailed the results of three turf samples analysed on the council's behalf).
73. SCOOT press release and *Forres Gazette*, 19 December 1990.
74. Sinclair, Fiona. Interview with Robert Allen, 1991.

75. Sinclair, Fiona. Interview with Robert Allen, 1991, DAFS. Letter to David Gerrard, 15 October 1990, Cowan, Sir Robert. Letter to F. Sinclair 21 December 1989, *Highland News*, 11 November 1989 and Nontox/Lanstar Inquiry Reporter's Letter, 26 April 1990, paragraph 52.
76. *Inverness Herald and Post*, 3 August 1990.
77. Sinclair, Fiona. Interview with Robert Allen 1991
78. "Planning reforms distrusted", *Glasgow Herald*, 8 November 1989
79. "Council anger over planning bombshell", *Inverness Courier*, 12 January 1990.
80. Nontox/Lanstar Inquiry Reporter's Letter, paragraph 153, 26 April 1990.
81. Nontox/Lanstar Inquiry Reporter's Letter, paragraph 154, 26 April 1990.
82. Sinclair, Fiona. Interview with Robert Allen, 1991.
83. "Nontox make an effort to change their dirty image", *Aberdeen Press and Journal*, 7 July 1990 and "Toxic waste firm opens its doors to the public", *Inverness Courier*, 10 July 1990.
84. "Rifkind's rulings create Highland storm", *Glasgow Herald*, 8 November 1990.
85. Highland Regional Council Planning Committee report, paragraph 6.1, 26 October 1988.
86. Proposed Modifications to HRC Structure Plan Review 1989, section 11, p. 26, 26 June 1990.
87. FOE (Scotland) 1990 – see also note 88.
88. Highland Region Structure Plan Review 1989 sections 11.1 and 11.2, "Hard Cash from Highland Rock", *People's Journal*, 24 February 1990 and "Highlands target mineral developers", *Scotsman*, 13 February 1991
89. "Lanstar is bought in £4m deal", *Manchester Evening News*, 14 November 1990.
90. Background notes on Charterhouse Venture Funds, from Broad Street Attitudes, 13 November 1990.
91. Sinclair, Fiona. Interview with Robert Allen, 1991
92. Hunter, Stewart. Phone call to F Sinclair 12 July 1991 and "Pollution Watchdogs: Fresh Fears", *Highland News*, 13 July 1991.
93. "Lanstar sells chemicals division", *Irlam and Cadishead Advertiser*, 27 June 1991.
94. Sinclair, Fiona. Interview with Robert Allen, 1991.
95. Simpson, Alan (on behalf of HRC Director of Law and Administration). Letter to Mrs Mary Cocker, 16 May 1991.
96. Sinclair, Fiona. Interview with Robert Allen, 1991.

Chapter 3

PONTYPOOL AND RECHEM
Fear and loathing in a Welsh valley

David Powell was awake early that summer morning. It was about seven o'clock, the beginning of another day in college. Attracted by the beautiful morning – the air was still, the sky a radiant blue – he walked to the north-west facing window which overlooked the end of the eastern valley in Gwent. The view was normally majestic, stretching from Newport in the south through Cwmbran and up past Pontypool – green hills, a sloping green valley. What he saw was hideous. Black smoke was rolling down the valley, forming a layer over the Cwmbran area. The smoke spewed upwards several hundred feet from a source behind a hill to the north and poured down the valley, home to approximately 50,000 people. David Powell gazed at the source on the near horizon. Something was wrong. A factory ablaze? An oil tanker in flames?[1]

He began to get dressed, wondered about the smoke. "I don't think it was routine for me to get up and go to the window, but I was particularly taken by what seemed a delightful morning." Less than two years earlier, after 17 years in industry, he had decided to become a schoolteacher. It was the summer of 1984. He was 36 years of age, married to Denise; his daughter Nikki was six, his son Christian four. The college where he was preparing to become a teacher was near his children's school. He had the time to enjoy breakfast with them and drive them to school on his way to college. Sometimes he would finish college early enough to collect them, at the end of their day. Occasionally he would meet Denise at lunchtime, collect the children from school and go on a picnic in the countryside. Life was great. Yet something about that smoke disturbed him. He went up the lane for a better view of the source. "It was a view to be repeated, over and over again, at night as well as the early morning. It was a view which, that summer, took myself and a friend to the gates of ReChem at midnight, to stop the smoke coming over our homes."[2] The Powell family had lived in the large three-storey house on the rise above Ponthir, a small South Wales village with a closely knit community, for five years. Built by C.W. Ulett, the product of forty years labour, Ulett House is aesthetically gothic, stone steps up to the central front door, large panelled windows, pock-marked brickwork. The house is a monument to its builder. The Powells still live there. "In the adversity I now face with ReChem I feel in good company living in a house which must have presented its builder with moments of great despair, but in the end a rare sense of achievement."[3]

"Farm puzzle of paralysed sheep flock" ran the headline.[4] Newport farmer Colin Haines didn't know what to make of it. Approximately 18 months earlier his livestock had begun to die. A Fleet Street paper sent a reporter to investigate. Colin and Gwen Haines told a strange story; one morning Colin Haines awoke, looked out his bedroom window to the meadow below and saw half his flock of 150 ewes standing motionless. They dressed and went out to the field. "One by one, Mr and Mrs Haines went up to the motionless sheep and pushed them over. And one by one, the sheep fell limply to the ground, their legs rigid. After a few minutes some of the ewes started kicking, then got back to their feet."[5] The mystery at Bullmore Farm, in Caerlon, Gwent, had puzzled the Haines family. They had recorded the deaths of 120 animals (50 to the end of 1983, another 70 during 1984), including 44 ewes. During September and October 1983 tests carried out by the Ministry of Agriculture Fisheries and Food's veterinary laboratories in Gloucester on Haines' cross-bred calves to determine the cause of the debility revealed that the animals' immune systems had been supressed. The calves had been suffering from pneumonia and emphysema. One beast, aged five months, had died and was examined by the MAFF lab. The vet recorded that the animal "had crusty lesions over much of the head and neck and scattered over the body. These lesions resembled ringworm but were more severe than usually seen, with inflammation showing at the centres". The post mortem showed that the animal had died in very poor condition and was under-sized for its age. No fat tissue was left on the animal and its muscles had begun to waste away. Normal veterinary examinations revealed few clues to the cause of death. "We must assume that the poor condition of this calf was due to severe pneumonia," recorded the vet.[6] A MAFF spokesman later said: "We have carried out investigations and we can find no major disease conditions on the farm. Animals do die from time to time on farms and often it is terribly difficult to be precise what caused death."[7] During the early months of 1984, in correspondence with the Welsh Office, Colin Haines was told that his lambs were infected by a soil borne organism which caused arthritis. He was also told that a test for immunosupression done on his calves was negative.[8] Poor ventilation in Haines' calf house, it was claimed, had caused the pneumonia and led to their deaths.[9] In July Haines' MP Mark Robinson was told by the Welsh Office that investigations into the problems on Bullmoor Farm did not produce any evidence that they were caused by local pollution. The Welsh Office added that their investigations had shown that the domestic water supply to the Haines farm was of poor bacteriological quality, but it was unlikely that this was the cause of the problems, and went on to say that Newport County Council had carried out analyses of the grass, soil and water supply on Bullmoor Farm.[10] The following month Haines received a letter from the Welsh Water Authority in response to his complaint that his livestock were reluctant to drink water from the stream running through his land. As a result of samples taken and analysed, the water board's environmental officer was able to state that the water quality of the stream was satisfactory and that pollutant levels were well below the limits recommended for agricultural use as indicated in the Environmental Protection Agency manual "Water Quality Criteria" (1972).[11]

The mystery deaths at Bullmoor Farm and surrounding farms in the Gwent area baffled the community. Turning down an appeal for an inquiry into the animal deaths and deformities, the Welsh Office stated that there was no evidence which would justify a formal inquiry. A Welsh Office scientific advisor had visited Bullmoor Farm. The Secretary of State for Wales, Nicholas Edwards, said he would review the case if more information provided "a case for a more detailed investigation". There was, said the Welsh Office, no evidence to link the problems with local industrial pollution.[12] The Haines' vet, Robert Stephenson, wasn't so sure. The Welsh Office had told Stephenson that following the visit to Bullmoor Farm it was their belief that liver fluke and mucosal disease were responsible for the losses. The Welsh Office added that they could find no scientific evidence to begin an investigation at government expense, and Stephenson was urged to pursue the probability that the losses were due to fairly common disease conditions.[13] When Stephenson learned that the State vet had not tested for pollutants he insisted that industrial pollution tests be made. Although the State vet's diagnosis revealed common causes of death, Stephenson said it left questions in his mind "as to why recognised treatments have not been successful":

A deficiency in the immune system is a possibility. The cause of a breakdown in the immune system, making the animals vulnerable to disease, would not be easy to identify. I would not dismiss the possibility of dioxins of one sort or another. I would like to see what chemicals are around, not only on the Haines' farm but generally in the area. If information can be made available on levels of toxicity and any known symptoms relating to it, then it should be done."[14]

Later that September, Friends of the Earth announced a report which stated there was an obvious pollution problem in the Pontypool basin.[15] "We are not apportioning blame on anyone. But we believe there is a problem and there is definite concern among the people in the area. Nothing can be done until all the information is available."[16] Brian Price, FoE pollution consultant since 1972 and author of the report, said: "I am convinced there is something dreadfully wrong in that area. It looks very much like chemical pollution."[17]

Between 1980 and 1984 four Pontypool babies were diagnosed with extremely rare eye deformities; one child had tiny eyes, two were born with only one eye, one was stillborn with no eyes. In October 1984 the Welsh Office agreed to begin an investigation into the numbers of babies born with deformities in Torfaen. Leo Abse, the Torfaen MP, was told that the figures for Torfaen Borough were higher than other areas in Gwent and higher than the Welsh average. The figures, for the years 1978-82, stated that 18.7 babies per thousand had been born deformed in Torfaen. The rate for Gwent was 13.9 per thousand. The figures for eye defects were different. The Pontypool children suffered from microphthalmia (no eyelids) and anophthalmia (lack of an eye), conditions which normally occur in one baby per year in a population of 10 million. "These eye abnormalities are extraordinarily rare. As a result there is little medical knowledge as to what causes them," Richard Collin, consultant eye specialist at Moorfields Hospital, London, told a reporter in 1984. The doctor added: "The incidence in these two areas makes one extremely concerned," in reference to similar eye defects in Bonnybridge.[18] (see p Chapter 5)

The parents of the children could see no reason why this should have happened to them. Abigail Bown, born with tiny eyes, had her sight saved when she was operated on eleven days after birth; skin grafts were used to give her proper eyelids. "When Abigail was born the doctors said it was hereditary, but we checked with our parents and they know of no case in the family. Abigail is short-sighted but she can still see," said her parents Andrea and Keith Bown in 1984. Buckley Thomas, born with no right eye and his left eye not fully formed, is blind and brain damaged and was the first child known to be born with eye deformities in the area. "When he was born he went through all kinds of tests. They could not find any reason for his deformity," his mother said in 1984.[19]

Anthony Jones, a consultant radiologist working in the mid and south Gwent areas, commented in a letter to Nicholas Edwards on the statistics published by the Welsh Office on congenital defects. Dr Jones, a specialist in the early detection of foetal abnormalities by ultrasound, suggested that the statistics were not necessarily indicative of the true incidence of abnormality because they did not take into account the abnormalities detected in pregnancy which resulted in terminations. "For example," he wrote, "in the Royal Gwent Hospital between 1 July 1983 and 30 June 1984 3,754 women registered for antenatal care and of those pregnancies two babies were born with central nervous system (CNS) abnormalities and nine pregnancies were terminated because of early detection of CNS abnormalities. In effect the trends of incidence of abnormality cannot be meaningfully assessed unless all factors are considered."[20]

On 24 December 1984, following the *Sunday Times* report on the eye deformities, ReChem director Robin Drewett said that they would close their Pontypool works if any links were found between its emissions and the malformations.[21] Abigail Bown's parents said they were prepared to use their daughter as a test case if evidence could be found to connect her eye deformity with local pollution.[22] ReChem responded to the "irresponsible allegations" that it was to blame for the abnormalities in children; Allan Woods, ReChem's community relations officer at the Pontypool works, said:

> Allegations have been made that four babies have been born with eye deformities within a 10 mile radius. Without knowing the individual family history it is impossible for ReChem to defend itself against such allegations. If there are parents of babies with eye deformities who really believe ReChem is to blame, I would like them to contact me and we will arrange for a top eye surgeon to carry out independent investigation.[23]

The Bowns said they "would have no part in that at all." Abigail, Keith Bown said, had been in the case of Moorfield's specialist, Richard Collin, since she was six weeks old. "I don't feel there is another specialist in the world who can give her better advice or treatment.

No one has more knowledge or experience."[24]

ReChem came to Wales in 1972, initially to dispose of industrial waste generated in South Wales and in the south-west of England. By the end of the seventies the company was able to offer its customers high-temperature incineration to dispose of difficult wastes, such as PCBs. The antagonism to ReChem from the local Panteg community began largely with opposition to the company's operations from environmentalists; this intensified throughout the eighties when it was learned that ReChem were burning PCBs. Monmouthshire County Council, the local regulatory authority in the early seventies, had no worries about Rechem in the beginning. ReChem had built their works in an industrial area, situated on the floor of the valley, surrounded by commercial and residential buildings. Other nearby housing was located at a higher level, in the hill above ReChem on the down wind side of the plant. Smells and smoke had been a problem from the beginning. "Since the plant was opened in 1974 it has been the source of continual, justifiable complaints from both residents and occupants of industrial premises, not only in the Panteg area but also from areas lying further afield," Torfaen Borough Council announced, sometime after ReChem's 15th anniversary in Pontypool, when it was clear to the council that a high-temperature incinerator sited at the bottom of a valley was a bad idea. ". . . the systems currently employed at the ReChem plant, and the radical alteration to its pattern of business operations since it opened, are now such that its present location is wholly inappropriate and incompatible for its local environment."[25] David Powell put it in human terms:

"Since it first lit its furnace in Pontypool, ReChem has been infamous for its smells and smoke. From the senseless site, in the bottom of a valley, emissions can drift sideways or be swept to the ground. Atmospheric temperature inversion often puts a "lid" on the valley and keeps the cloud accumulating. For nearby residents life can be a misery, with closed windows and life indoors the only counter measure. Painful eyes, sore noses and throats, nausea and respiratory difficulties have been frequent testimonies of local people. Mouth ulcers and headaches are my own direct experiences of being in the ReChem wind. You don't need to be a toxicologist to detect that something is wrong."[26]

Something *was* wrong in Pontypool, but it would take government until 1990 to take action. The livestock deaths, the abnormalities in local-born children and the fear of dioxin contamination highlighted in

1984 brought the issue to the attention of the world's media; for local people the fear and loathing continued.

ReChem, however, was not prepared to be the scapegoat. In December 1985 the company was reconstructed when its management team, with assistance from the Charterhouse Venture Fund, bought out the parent company, British Electric Traction (BET), in a £1.8 million deal. Until December 1990, when ReChem was merged with the Shanks and McEwan Group, the company became Britain's leading toxic waste disposal operator with two of Britain's four high-temperature incinerators; profits increased almost ten-fold from £641,000 in 1985.[27] It was able to achieve this success by staying ahead of its competitors and by keeping its critics at bay. "Critical exposure of the wrongs of ReChem has been suppressed by legal intimidation," David Powell wrote in 1988:

> ReChem has barricaded itself behind the UK's libel laws, firing threats at anyone who could damage the company's commercial image. The media tends to draw back from the more serious aspects of the issue. The BBC itself is quite afraid to engage in critical coverage of ReChem and often ignores important developments in evidence against the company. Friends of the Earth, Greenpeace, most newspapers, TV and radio companies have either been warned off or been sued.[28]

The concern in the community before ReChem built their incineration works was largely theoretical. No one really knew what the dangers of incineration were. Technical information about PCBs and the effect of dioxins and dibenzofurans on human health was thin on the ground and then, as now, not enough evidence has been gathered to prove a case in court. (PCBs are organohalogens, the collective name for the bonding of carbon and chlorine, one of the synthetic products of the petrochemical industry. Because organohalogens are synthetic and are stable compounds, the organisms that break down natural wastes do not have much effect on them, with the result that humans and animals can accumulate chemical compounds like PCBs in their bodies.) The burning of PCB waste was not included in ReChem's initial planning application. During the mid-to-late seventies when the communities began to experience eye, throat and stomach aliments no one – local GPs, scientists, environmental health workers – could attribute the cause to local pollution. The days of toxicological tests and epidemiological studies were in the future (and to a certain degree still are). Complaints about localized pollution in the Pontypool area increased throughout the seventies until it became evident that the local authority did not have the power to deal with the problem. If the source of the pollution – the ill-health in local people, the livestock deaths – was ReChem, could Torfaen council prove it? Could the council tell ReChem to go away? David Powell: "It seemed to me that people believed that ReChem was a rather small scale issue which could be dealt with by the existing powers of the local authority. They didn't comprehend that to move ReChem was really like moving mountains. The powers needed would be much greater than the local authority."[29] The Panteg Environmental Protection Group (PEPA) had been the only significant thorn in ReChem's side in the early days. Retrospective criticism of PEPA must be placed in context, that the group were ignorant of the problems of incineration and were unable to put any real pressure on the company to account for its activities. PEPA and ReChem met frequently to discuss complaints which were also recorded by the council, but no progress was made. The complaints intensified as the community insisted that ReChem's fumes were affecting them, but there was no evidence to prove this. By 1984 ReChem was insisting that they were not emitting PCBs from their plant. ReChem, aristocrats of the toxic waste trade, saw no threat from their plebeian critics. The need for a government sponsored public enquiry into industrial pollution in the Pontypool area was paramount.

The local campaigners already had ten years experience when David Powell decided to find out more about the source of the black smoke he had seen from his bedroom window in the summer of 1984. It didn't take him long to discover that the source was ReChem and he approached the company, the council, the Pollution Inspectorate but came away dissatisfied:

> I certainly felt there was an attempt to pull the wool over my eyes about the real activity and of the regulatory procedure. I assumed I wasn't going to get the information I wanted through the proper channels so I decided to seek that information through less official means. I set off to contact as many people as I could who knew about ReChem, incineration and toxic waste; they were very helpful, also in establishing other contacts.
> I also contacted the people in what was then the major anti-ReChem campaigning group – PEPA. We helped each other and together we became involved with the Scottish people. I decided because I'm geographically distant from them we would be better off if my

campaigning strategy was slightly different, so Stop Toxic Emissions Action Movement (STEAM) was formed. There were two things that we tried to do: to increase the amount of information that was available to the public and to local authorities, and to increase our own awareness of the whole issue. These two things went side by side and included the 24-hour-a-day week-long monitoring of ReChem which we called the "toxic watch". That proved to be very valuble in information gathering and publicizing the issue. We did find that a lot of people knew of ReChem, where it was and obviously knew that it smelled and smoked but otherwise were very hazy about the nature of ReChem's activities. They didn't know that it was dealing with all manner of different wastes, . . . from countries all over the globe. In fact it's fair to say that there were very few people in south Wales who were aware that ReChem were burning PCBs.

ReChem also gave us some information. If you asked them whether they were burning PCBs they were obliged to answer. They also answered our question about the burning of radioactive waste; they admitted they were burning radioactive waste. They were less positive about the myriad of different materials they were burning and what was being burned with what. The local authority was reasonably helpful in identifying some aspects of the regulatory procedures. The Industrial Air Pollution Inspectorate and the Factories Inspectorate were of almost no assistance. The main avenues of progress were from the Scottish people, from people overseas. Bit by bit we got a picture of what ReChem was doing, what materials it was dealing with and how it was doing it.[30]

Gathering information on the pollution in the Pontypool area was to become a laborious process for the community; for groups like PEPA and STEAM and later, Mothers and Children Against Toxic Waste (MACATW); for people like the parents of the children with eye defects; for the farmers who had lost livestock; for activists like David Powell. In 1988 in his appeal to the Canadian people to ask them to tell their government to stop its exportation of PCB waste he wrote:

We are victims of the intransigence of a policy of non-interference with market forces. We need help from outside. We need your help. If you take the easy way out by ditching your waste . . . you will not assist the advancement of a sensible and secure system of hazardous waste disposal and ultimately clean technologies to reduce waste at source."[31]

David Powell's impassioned plea was written four years after ReChem's Bonnybridge works had closed down and long after it was acknowledged that the Pontypool works had the only high-temperature incinerator in Britain capable of taking solid PCB waste. ReChem had believed, since 1980, that it could take PCB waste and destroy it effectively, operating at temperatures of between 1000 and 1200°C. This, along with other criteria, seemed to enable ReChem, as one reporter paraphrased it, to take dangerous chemicals such as PCBs and break them down into harmless by-products.[32] Criticism of ReChem's incineration process by the environmental organization Greenpeace brought a 12 page response from the company. In the document ReChem stated in response to specific criticisms from Greenpeace:

Test burns on ReChem International's incinerators were carried out under normal operating conditions, and not under "ideal conditions". Test burns are also supplemented by a regular sampling scheme to ensure constant monitoring. The plants are built, designed and operated with substansial margins for human and mechanical error. The incinerator temperatures are monitored and recorded all day and night and operating conditions are strictly observed.[33]

Harwell (laboratories) have completed a through and rigorous investigation of our incinerators. They pronounced our system safe and efficient for the destruction of PCBs, which confirmed our own sampling results. We are able to maintain the conditions that lead to this pronouncement at all times, and there is no reason or evidence to suggest that the incinerator is not adequately controlled.[34]

The allegations of economic injury inflicted on the community around the plant are unfair, unjustified and unfounded. Not only has no "casual connection" been "officially established" it has been repeatedly "officially" denied. THERE IS NO EVIDENCE AT ALL, CIRCUMSTANTIAL OR OTHERWISE, THAT RECHEM INTERNATIONAL IS UNABLE TO DEAL WITH PCBs AND OTHER CHLORINATED HYDROCARBONS. IN FACT THE EVIDENCE IS TO THE CONTRARY. [The emphasis is ReChem's.][35]

We became aware of concern that intermediate products such as dioxons and dibenzofurans might be formed during incineration and not destroyed, and we took responsible action and stopped burning PCBs until our plants were demonstrated to be efficient by an independent body. A full and final report came through from Harwell in 1980. That gave us the "all clear", stating categorically that we could destroy PCBs effectively."[36]

The community in Pontypool wasn't as easily convinced. Something, as the community had asserted for over a decade, was wrong. Yet no evidence existed to show that PCBs were not properly destroyed in incinerators like ReChem's. Test burns initiated by government, and carried out by the Industrial Air Pollution Inspectorate, revealed no detectable emissions of dioxins and dibenzofurans.[37] This wasn't enough to persuade the community that incineration was safe. A detailed survey of the attitudes of approximately eight hundred Pontypool residents found that three quarters were opposed to ReChem.[38] The official belief reinforced "repeated statements by the regulatory authorities and plant operators that these facilities are operating safely and efficiently",[39] but had little or no impact on the community's attitudes. As Women's Environmental Network (WEN) scientist Ann Link would wonder: "I have been wading through a lot of published scientific stuff on toxicity and evidence of effects, and feel that the so-called "anecdotal" evidence is screaming through all the distanced scientific papers."[40] "PCBs are a particular problem. There is no argument about their danger to the environment," David Powell asserted when he attempted to place the debate in the context of the environment and the dangers to human and animal health:

> There is no argument about the need not to produce any more PCBs – that has been accepted. . . . There are two things I question; one is the need to bring them out of use quickly and whether that action does more damage to environment than leaving them in use under controlled conditions. . . . The other is the myth that by shoving them out of one country into another you are automatically solving the problem. . . . For the people in the area where the PCBs are going to, you're creating a problem.[41]

Throughout the eighties the battle between the communities and ReChem raged virtually unabated. The cordial tripartite meetings between PEPA, Torfaen council and ReChem had been a refrain of the seventies. Yet nothing had been achieved despite the depth of feeling. Public meetings during the eighties appeared to alter the mood. "Quit our valley," was the frequent cry from the community. ReChem defiantly said they would remain, and even announced they would expand their operation. Following one meeting in November 1984 ReChem managing director Malcolm Lee told a local paper that the company had been directed to the site by Monmouthshire County Council.[42] Torfaen MP Leo Abse argued that ReChem should bear the sole responsibility for the choice of site. "The planning decision was made by an unsuspecting council hungry for new employment in the area."[43]

STEAM, the newly formed anti-ReChem group, wanted to adopt an approach of direct action. One of the group's first initiatives brought them into conflict with the established mode of life in Pontypool. STEAM announced they were planning an alternative prospectus for companies wishing to locate in the Llantarnam Industrial Estate in Cwmbran. The group argued that the site could not provide clean air for incoming firms and their workers because of the alleged danger of emissions from ReChem. The local media responded with vitriolic editorials.

> If we see any real evidence that ReChem is harming the environment and local health, we shall lead the protest march. But there is no evidence to support the view that the air around Panteg is unfit for other industry or unfit for workers and their families to breathe. There is ample evidence that unemployment is damaging to health. It is far, far mores destructive to the environment than ReChem. . . . We do not need pressure groups who turn valid concern into what seems like luddite blackmail. Life, the environment and our community are too dear to be used as an emotive pawn by those who seem to see no further than the ReChem chimney."[44]

Emotive or not, there was still something wrong in the valley, but to establish that it was pollution by dioxins would be costly. Torfaen council's environmental health officers had taken six soil samples from the area surrounding ReChem; laboratory tests on the samples cost £2,400; a regular monitoring programme would cost ratepayers £50,000 a year. "The finding of dioxin is only the start," David Thomas, the chief environmental officer, said: "Proving where it came from is another thing." It was "totally unfair", he added, that a borough council should have to take on such an expensive programme. The government should make a contribution.[45] Failing to get any action from the government, the anti-ReChem groups announced they would hold their own inquiry into the disposal of PCBs at ReChem.

The "toxic watch" the groups had placed at the ReChem gates had an immediate response, but the joy of the campaigners didn't last. Following the second toxic watch between 6am 28 May and 6pm 31 May 1985 STEAM concluded:

> It works! In relation to smells and smoke the surrounding area has had one of its best ever weeks. One of the aims of the TOXIC WATCH was "to make ReChem cleaner for a week". We have clearly succeeded in this. Our presence has obviously influenced ReChem's performance. Our TOXIC WATCH has done more to improve emissions that the official regulatory authrities have. Organised community monitoring will continue, in a different form, from now on, indefinitely.
> We have stretched our knowledge of ReChem's activities far beyond that obtainable through the company or other official sources. . . . The regulatory authorities are, at most times, complicit with ReChem in their evasiveness, unwillingness and something their inability to answer

sensible and genuine questions of concern.

With the knowledge we have accumulated about the materials, the process, the systems of work and the shortcomings of all of these, we are now in a position to expand our existing recommendations. We are writing a second letter to Local Authorities, Government and the business interests requesting the implementation of these additional recommendations. They are recommendations which we acknowledge translate into significant increase in cost. Up to now the price for hazard and nuisance has been paid by the community. This is not longer acceptable or negotiable.[46]

The toxic watch had consisted of round the clock observations and systematic recording of materials in, materials out, and stack emissions. Nearly a hundred volunteers recorded 714 separate observations which included vehicles off-loading, movement of materials, storage and incinerator loading. STEAM compiled a document based on their toxic watch and sent it to the Secretary of State for Wales, the Secretary of State for Trade and Industry, the Secretary of State for the Environment, BET, ReChem, the Chemical Industries Association and to Torfaen council. Copies were sent to several Westminster MPs, including the Prime Minister, to MEPs Llew Smith and Alex Falconer, and to Gwent County Council and the HSE.[47]

It had been a good Spring for the groups. During April they had heard from several experts who spoke on the need for ReChem to improve their operation and their handling of chemicals like PCBs. Dr Edward Kleppinger, a US consultant who spoke on incinerator technology, stressed that it was necessary for citizens' groups to monitor ReChem's operations "100 per cent" because there had been failures in regulatory action. Recent "housekeeping" at ReChem would not have been done if it had not been for the action of the protestors, he said.[48]

In May the icing on the cake, it seemed, was the result of investigations by scientists working for a BBC programme. A partly burned capacitor containing PCBs had been discovered on a council waste tip. The belief was that the capacitor had come from ReChem. The company denied this. David Powell alleged he had seen capacitors dumped by Biffa lorries. The campaigners had, with the help of the BBC, made their first major mistake. David Powell:

Back in 1985 the BBC *Newsnight* team became involved in two programmes on ReChem and toxic waste. Because they were sued by ReChem over the first programme the second was never produced. The BBC, after speaking to some American scientists, took the view that if someone wanted to discover evidence of contamination as cheaply and conveniently as possible then it wasn't necessary to look in the output from ReChem's stack or in the output from ReChem's liquid effluent; the key to it would be to look at the solid residue that ReChem had produced. The American scientists said that if you find high levels of PCBs in the solid residue of burned material then you can guarantee there is a high level of dioxins and furans there, because the stuff hasn't been properly burned.

The *Newsnight* team went to the tip – the local authority tip, which has an industrial section used mainly by ReChem; I actually saw lorries tipping whole capacitors which didn't look as though they had been burned much. They certainly weren't broken up and ashy, they were solid lumps of stuff. *Newsnight* took some of the burnt capacitor material from a typical ReChem pile which presumably came out from one of ReChem's skips. *Newsnight* had the stuff analysed and found in one of the samples high levels of PCBs. It seemed to be all over and done with; we had the evidence that ReChem was producing contamination.

ReChem got off in the simplest of ways. They said: "You can't prove it was our material." ReChem were the only company incinerating capacitors in the UK; that material was from a burned or partly burned capacitor. ReChem, quite rightly, supposed that if it was to be pinned on them it had to be proved it was their material. Nobody could do that.[49] (In 1987 the BBC agreed to apologize to ReChem.)

The ReChem management's ability to interpret the letter of the law would also give Torfaen councillors constant headaches. The council appeared to be powerless in their attempt to impose some sort of control over the company's activities. This inevitably led to antagonism between them. "It is with some regret that my council has faced what has become a campaign of abuse directed against (it), its members and officers by (ReChem)," councillor Brian Smith, leader of Torfaen council, told the Welsh Affairs committee in December 1989. He added that ReChem believed Torfaen council was "engaged in an unprincipled campaign of hostility aimed against the company with the intention of causing it to close down". ReChem, Malcolm Lee told *Director* magazine, had been the victim not only of hysteria, exaggeration and misguied idealism but of lies told for political gain. "What happens if you are, say, a local politician and want to make a name for yourself, you have only got to go scaremongering at the weekend and you get the front page – it is guaranteed! We have become a *cause célèbre* for the left-wing politicans down there." Lee's comments were quoted to the committee by Smith who said:

I have no intention or wish to become engaged in a slanging match with ReChem. However, I think it perfectly proper for me to draw these remarks to your attention and to deny each and every one of them as being baseless, without any foundation in fact and untrue. My council's position has been quite clear and quite consistent. They believe that there are good reasons for the Secretary of State to hold a public enquiry and have consistently stated that if there is a public enquiry my council will accept the findings.[50]

The real problem Torfaen council faced was the entanglement of regional environmental legislation over whether the council had the legislative power to monitor ReChem's activities and seek prosecutions. When Torfaen decided to prosecute ReChem for causing a nuisance contrary to the conditions of their waste disposal licence and under the Control of Pollution Act, they discovered that the law was not on their side. Judge Garland ruled: "As a matter of very technical law the council has no statutory power to impose the conditions ReChem were alleged to have breached and since the whole case was about the breach the whole house of cards comes tumbling down."[51] ReChem's original licence had included a clause which prohibited the company from causing a public nuisance.

Both the council's and the company's interpretation of the environmental legislation is a clue to why Judge Garland declared Torfaen Borough Council's attempts to enforce the nuisance clause *ultra vires*. Under the Control of Pollution Act (1974) Torfaen Council must issue a licence and impose conditions to companies like ReChem. This piece of legislation also makes Torfaen council a Waste Disposal Authority. Under the conditions of the licence the WDA should be able to regulate companies such as ReChem so that they do not cause a danger to health, do not contaminate water and do not cause serious damage to the amenities of the neighbourhood in which they operate. The WDA, however, must liaise with the HMIP and the Water Authority, and it has been Torfaen council's contention that, in relation to ReChem, this tripartite arrangement has been compromised.

> In practice it can be demonstrated that the council's activities to safeguard its inhabitants is, under certain circumstances, impeded rather than aided by these other statutory organizations.
>
> With regard to the Welsh Water Authority and the council's requirement to ensure the safety of water supplies, it is quite obvious that it is necessary to know what consent levels have been granted to ReChem for its discharges into public sewers. This information is not available to the council because "the consent agreement is a confidential matter between the company and the Water Authority".

The alleged commercial secrecy of these sewer discharges makes a nonsense of supposedly environmental conscious organizations such as the Water Authority and places a serious legal impediment in the path of the council's attempts to carry out its full role as a Waste Disposal Authority.

With regard to HMIP the ReChem incinerator is a registered process under the Alkali Acts and as such the HMIP have responsibility not only for the standards of incinerator operation but also for the ancillary processes such as material storage and handling, which includes the "transformer cutting and draining" operations. Because of this, the waste disposal licence is unable to include specifications relating to the technical running and control of these operations and can in no way cross, impose upon, or even improve what may be contained in HMIP's "Best Practicable Means" document. This situation is made even worse by the fact that there is apparent confusion amongst the ranks of HMIP as to who does what.

The recently revised document "Waste Management Paper No 4", which deals with waste disposal licences, and which was largely written by HMIP, clearly states that body will decide what materials may be incinerated in such plants as ReChem. Despite this clear statement, a recent letter from HMIP to the council (written at the time of issue of the revised Waste Management Paper) states that it is the council who decides what materials can and cannot be incinerated.

Furthermore under the Alkali Acts, HMIP can only lay down emission standards for aerial emissions for acidic gases and particulate matter. There is obviously no control measure available for laying down standards for PCB, dioxin and furan emissions, whether from the chimney stack or any other parts of the process such as the cooling towers, which emit high volumes of odourous water vapour from the cooling of contaminated gas cleaning water. There are obviously great gaps in available control measures which could be easily closed by the extension of control via the waste disposal licensing system and which would readily come to light in the process of a public enquiry.[52]

This confusion, Torfaen council claimed, "led to a great deal of public frustration and lack of confidence" as the majority of people tended to direct their first line of complaint at their own council. "Since most complaints relate to HMIP controlled operations complainants have to be redirected to the appropriate authority."[53]

Since all matters covered by HMIP take precedence over the local controls available via the Waste Disposal licence issued by Torfaen Borough Council (as confirmed by the Waste Management Paper no 4) it is again obvious that the authority "on the ground", in the best and immediate position to deal with such matters is precluded from doing so, and illustrates again the fact that proper licensing control over such plants is impeded. Further proof of this statement is brought by way of reference to the council's unsuccessful prosecution of the company for creating nuisance to the inhabitants of the area. As the Secretary of State will no doubt be aware the Court of Appeal has recently held that a Waste Disposal Authority cannot include in its licensing conditions clauses which prevent such operations causing nuisance. It would seem from this decision that whilst such plants should not be allowed to endanger health they may have little regard to the quality of life in their neighbourhood since frequent nuisance cannot be controlled by the licensing system.[54]

ReChem, in a letter to Torfaen council, argued that the council had failed to implement the legislation as it stands. In a point by point analysis ReChem spelled out what it believed were the inadequacies of the council's argument, how, in the company's words, it "exercised its duties and powers under this legislation". ReChem referred to the Control of Pollution Act (1974), the Alkali and Works Regulation Act (1906) and the Public Health Act (1936). On sections 5 and 6 of the CPA ReChem stated there is no condition on the licence in respect of atmospheric emissions and that the council had never served notice under section 7, nor sought to amend the licence. The company also stated that the council had not taken any action under sections 9, 79 and 80 of the CPA and on section 93, which gives the council power to serve notice requiring information it needs to carry out the functions imposed upon it by the act, ReChem stated: "Torfaen Borough Council has served only one notice under this section which was considered *ultra vires* because it sought information relating to the statutory function of another body (ie HMIP) who had already been given the information by ReChem." The company also stated that the council had never applied section 22 of the Alkali Act, which allows the council to make a complaint to the Secretary of State who in turn may take action. Under sections 91 to 94 of the Public Health Act, which refer to nuisance, ReChem stated that the council had taken no action.

It is therefore patently obvious that the British government had enacted legislation which is more than adequate for Torfaen Borough Council to fulfil both its legal and moral obligations. That it has singularly failed to use its powers to fulfil its duties is a matter for serious concern, clouds the real issues and casts considerable doubt on the alleged reason for calling for a Public Inquiry when the facts are already available and the means to obtain these facts contained in the statutes. Torfaen Borough

Council's failure and abdication is not sufficent reason for an Inquiry into ReChem's activity.[55]

In a lengthy answer to the Welsh Affairs Committee the principal legal assistant to Torfaen council, Chris Tindall, said that the points in the ReChem letter were wrong; by including them in their evidence the company had placed several incorrect statements of law before the committee. Concerning the revision of ReChem's Pontypool licence, Tindall said there was no point preparing a new licence until the results of the company's appeal against the refusal of a licence for their Fawley works had been heard, "because, plainly, the Fawley plant licence will be a precedent". He said that sections 7 and 9 of the CPA "would be ideal for us to find some trumped-up excuse to revoke their licence"; the council, he stressed, has not done that but had tried to deal with ReChem with the resources available to it. Section 9, Tindall said, imposes a duty on a waste disposal authority to take the steps necessary to ensure that pollution, including the pollution of water, does not take place. Tindall went on to state that the dispute between the council and the company was complicated:

. . . Torfaen Borough Council are a waste disposal authority, as defined by section 30 of the Control of Pollution Act. They are a district council. Most waste disposal authorities in England are county councils. In Wales it is a district council function. We also happen to be a local authority as defined in a later section of the Act – that section which deals with air pollution – for the purposes of taking action under sections 79 and 80. But we have maintained throughout that we have the powers of a waste disposal authority and have the wider powers under section 9 to carry out the sort of tests we have been doing. And in fact in September of this year (1989), ReChem issued proceedings against the council (a) seeking an injunction prohibiting the further publication of a document; and (b) – and inherently tied up with those – saying that there was a judicial review required so that the courts could decide on ReChem's contention that Torfaen Borough Council, in carrying out the tests, were acting illegally because they were limited to the powers of a local authority, a district council, and not a waste disposal authority seen in the terms of an English country council. There is, therefore, this split between the concept of a waste disposal authority's powers, which, as I say, is usually in England a county council; and a local authority, usually in Wales a district council.[56]

ReChem had, in fact, withdrawn its case on section 9, and the council had taken action under the section. Tindall explained that the reason the council had acted under section 9 was because "section 79 gives you powers to investigate but does not actually give you powers to do anything. A district council cannot take any action whereas a waste authority can. That is the argument". Section 80 of the CPA gave the council power to serve notices for information on emissions. Tindall: "Our major concern in relation to these matters is that we believe the current problems arise from emissions from the furnace or certain activities that are taking place in the open air – cutting up, opening of barrels, etcetera. They are the emissions that we wish to investigate." (Tindall added that the council's position in relation to the Alkali and Public Health acts required the authority of the Secretary of State and that he had indicated that he was not willing to grant permission for proceedings to be started against ReChem.)[57]

There was, of course, more to the council's and to ReChem's inter-pretation of the existing legislation than met the eye. It was obvious to the campaigners that the legislation did not enable the local authority to take specific action against any polluting company unless proof could be obtained and that was the problem facing Torfaen council. The existing legislation simply required companies like ReChem to self-monitor their activities and report retrospectively to HMIP. As far as the campaigners could see, that would not implicate ReChem in any illegal activity. ReChem had stated that Torfaen council had refused to join a liaison committee similar to those which functioned between the company and the respective local authorities in Fawley and Bonnybridge. The "stumbling blocks" as Tindall put it were: "Firstly as well as general information, we have indicated that we would like information about specific incidents as and when they occur. We have indicated that there is no point in being a member of a liaison committee that cannot receive up-to-date information if a particular incident occurs."[58] ReChem, however, did not see why it should provide this information under the council's terms. Their reasoning became apparent in a letter from Malcolm Lee to the council on 26 May 1989.

[I] felt unable to consider joining a liaison committee that dealt with specific incidents within six months of them happening – because that is the time recognized as being the time in which prosecutions can be brought in most circumstances. I note your concern about the discussion of specific incidents. I am advised that it would be contrary to the rules of natural justice for ReChem to be expected to supply information to be used on the basis for a possible prosecution or cross-examination. Although I want to co-operate and assist, because we have nothing to hide, nevertheless I am conscious that I would be in breach of my duty to ReChem shareholders if I voluntarily surrendered such a basic legal right which has been enshrined in the British consitution since the time of Magna Carta.[59]

It appeared that it was not the council, it was ReChem who had reservations about a liaison committee. The council, Brian Smith said, was "prepared, even at this stage, to enter into the possibility of a liaison commitee provided that the company puts all of that information on the table, that nothing is hidden or swept under the carpet and that there is no interference by the company as to who should, or should not, be members of such a committee". Smith added "If I go on a committee like that . . . I want to be protected and to receive an undertaking from the company that anyone joining the committee will be free from threats or actual prosecutions, that they will be able to express open and honest opinions.[60]

David Powell summarized the problem:

It seemed that the major power Torfaen Borough Council had over ReChem was contained in that licence. The conditions of the licence are not really specific and they certainly don't relate to the amount of contamination permitted and so on. One of the clauses is about nuisance.

After many years of residents complaining about the smell and the smoke from ReChem, Torfaen council actually took them to court. They brought five specific allegations of ReChem causing a nuisance which occurred in 1985; they seemed to be fairly solid cases with very good testimony – one involving a woman who had been physically sick and her doctor had been present during one the of times she was sick. He agreed that it was connected with the smell and phoned ReChem who admitted they had a problem with some material, so it seemed quite a straightforward case which the local authority would win. After several days it was suddenly thrown out of court with the unbelievable idea that Torfaen council had no entitlement to take ReChem to court . . . in connection with such a licensing condition. In my view the licence wasn't worth the paper it was written on. The council appealed against the decision and the appeal was rejected.

Torfaen council has, I know, considered withdrawing ReChem's licence and the public have pressurized it to do that but the council have rejected that idea in the belief that (ReChem) would carry on without a licence and, secondly, without a licence Torfaen would have less than the tiny bit of control which they can exert at the moment.[61]

The controversy over ReChem's toxic waste imports became, during the late eighties, a rallying point for the company's critics. The campaigning groups vociferously opposed the importation of PCB wastes from Australia, the Netherlands, New Zealand and Canada.

They tenaciously lobbied foreign civil servants, drafting letters to embassies, organizing meetings and pickets in an attempt to prevent the waste coming to South Wales. It was important for the groups to concentrate the campaign on the international toxic waste trade. ReChem had exploited a marke; Malcolm Lee claimed "people were knocking on our door"; a former ReChem plant manager said the company had agressively pursued PCB waste. "We had the competence and the ability to handle (PCB) waste and therefore we attracted it. . . . It is not a question of their shipping it over because we will take it; it is very much that we went out and sought it as a business enterprise." PCBs are a high-revenue earner, said the former employee. Lee said the market "existed because of pressures [due to] the growing awareness of the problems of PCBs. We had the facilities. Most other countries did not".[62] David Powell did not see why ReChem should take this waste.

> The problem of toxic waste will be solved by people taking responsibility for the waste they produce. As long as one person, one community, one factory, one country can pass its waste onto somebody else then there is always the possibility of turning a blind eye to what goes on with that waste. We've got to close down the outlets for waste and make waste disposal as local as possible.[63]

Until that happened the campaigning groups had to keep the pressure on. The Australian trade unions blacked the movement of the PCB waste. The Dutch government said it wanted controlled and safe incineration of PCBs to prevent further pollution. If or when the British government decided that safe disposal at Pontypool was no longer possible then the Dutch authorities would endeavour to prevent the further export of wastes to South Wales. As 13 tonnes of PCB waste left Wellington, STEAM asked the New Zealand High Commission in London to suspend toxic waste exports until the effects of its incineration had been assessed. After the Karin B saga, (see p. 178), the relevation that PCB contaminated waste bound for ReChem had been carried on an Air Canada passenger flight sparked off a new campaign by the anti-ReChem groups that brought a remarkable success. (Between November 1988 and June 1989 Air Canada shipped 27 tonnes of toxic waste in three separate consignments in passenger flights from Toronto to London.)

The campaigners realized that if they were to make any impact on ReChem's activities they would have to stop the company's lucrative importation of toxic waste, and they would have to do this on their own. The Welsh Office had informed David Powell that waste to

ReChem was properly dealt with and there was no reason to restrict the trade.[64] Chris Patten, the Secretary of State for the Environment, said the movement of all wastes were subject to strict controls. Only five per cent of waste destroyed in Britain came from other countries. "We have a first class technology for dealing with wastes," he said. "It would be really disastrous for the world if these wastes were to be dealt with less safely in other developed countries or simply dumped in the Third World."[65]

In October 1989, a year after David Powell had petitioned the Canadian High Commissioner in London, the Canadian Council of Ministers of the Environment agreed to halt the exportation of PCB wastes for destruction in other countries. Exports of PCBs to the UK from Canada would "not be allowed due to the uncertainty of their acceptance". Lucien Bouchard, the Canadian minister for the environment, told MEP Llewellyn Smith that Canada was "now examining its legislative initiatives to better control the movement of hazardous wastes beyond its borders to protect the global environment".[66]

The campaigners had achieved a major victory. ReChem would continue to import foreign PCB waste but Canada, which had been the second highest exporter of PCBs to South Wales between July 1989 and May 1990 (after Sweden, which exported 640 tonnes), had withdrawn.

There had been other successes. Vigils outside ReChem's Pontypool works, protest marches and away days to Liverpool and Newport docks had forced the company to respond to the so-called PCB crisis with large adverts in the press. The Liverpool dockers had agreed not to handle toxic waste, although 140 tonnes of PCB waste had been transported through Newport docks between April and June 1989. During the week that the Canadian government decided to halt the exportation of PCB waste the port authority in Pembroke docks announced they would not handle PCBs from Canada.

Throughout 1989 and 1990 Sweden, Canada, Italy, Switzerland, Belgium, Australia, Spain, the Netherlands and (West) Germany had shipped a total of 2,389 tonnes of PCB waste to ReChem. The campaigners had failed to stop the waste from Switzerland and the Netherlands during earlier campaigns. Following the success of the campaign against the Canadian waste, the groups stepped up their protests and targeted the Netherlands, Sweden, Switzerland, Italy, Belgium and Austria. In August 1989 the groups pledged their support to David Powell who had been threatened with legal action by ReChem over remarks he had made about the Canadian waste. A public meeting and a march to ReChem reflected the anger in the community over what people in South Wales saw as intimidation and victimization.

Following representations to the ambassadors of Belgium, Italy and Sweden the groups led a protest to the embassies of Belgium and Sweden. The scientific attaché at the Italian embassy said his country had begun the construction of plants to enable them to cope with their own waste problem. Reflecting on the discussions with representatives of the Belgian and Swedish embassies David Powell remarked: "We are constantly reminded of Britain's apparent willingness to accept such material. We feel sad that so many of our ports are carrying on this trade and although we appreciate the difficulty in saying "no" we would like to see some indication that workers at the ports are not happy with the position".[67]

The Belgian media, which had been alerted to the embassy protests, had given the campaign prominent and extensive coverage but it wasn't enough to stop the toxic waste trade. "There is, however, nothing the Belgian government can do, as long as the import of toxic waste from private companies by ReChem is absolutely legal and that ReChem is a licensed company," the Belgian ambassador told Sarah Preece of MACATW. He stressed that the campaigners' petition and documents had been sent to the competent authorities in Belgium.[68] Sarah Preece had invited the ambassadors of the Netherlands, Switzerland, Italy, Sweden, Australia, Austria, (West) Germany and Belgium to Pontypool "for a presentation of our position in this important issue", after the campaigning groups had led a deputation of 40 people to the Dutch, Swiss, Swedish and Italian embassies in July. Attachés at the Dutch, Swiss and Swedish embassies told the protestors that they were doing nothing illegal. The Italians turned the protestors away. The Dutch, in response to Sarah Preece's invitation, said she "must convince the Inspectorate of Pollution and not us, that the present situation is not acceptable".

The campaigning groups had used moral pressure from the community to stop the ships bringing PCB waste from Canada. But their major argument against the toxic waste trade was to persuade developed countries to take responsibility for their own waste.

When the government finally announced in August 1990 that it would commission an independent survey to establish levels of PCB contamination in the general vicinity of ReChem's Pontypool works, the Secretary of State for Wales, David Hunt, referred to uncertainties over the tests carried out by Torfaen council and by the Welsh Office. He announced in January 1991 that the University of East Anglia would carry out the investigation of PCB levels and added: ". . . one of the factors which influenced me greatly was the level of public concern in the area. In the conduct of this survey I am anxious that those in the locality should have an opportunity to make their voices heard."[69]

Yet the community had already made their voices heard. The Caldicott family, who live at Pontyfelin House across the road from ReChem's works, complained initially about odours and fumes, and subsequently about high levels of PCB contamination on their land. Ken and Shirley Caldicott had been living at Pontyfelin House for seven months when a thick beige cloud, with an aromatic smell, engulfed their children who were playing in the garden. Ken Caldicott phoned the Environmental Health office of the council and ReChem. The incident, on 31 July 1988, was noted by Ken Caldicott in the diary he had kept on ReChem emissions since December 1987. A few days later the council took samples of the grass. On 20 February 1989 the area around ReChem was showered with charred aluminium foil. Some of the foil landed in the Caldicott's garden. Again the local authority was contacted and samples taken. Ken Caldicott noted the incident in his diary. Subsequent analysis by the council of the grass, the foil and eggs laid by the Caldicott's pet ducks revealed high levels of PCB contamination. ReChem questioned the findings. Ken Caldicott noted that the analysis of the duck eggs showed PCB levels six times that recommended by the Canadian authorities. This was backed up by further analysis. Then something strange happened at Pontyfelin House. "In a six week period up to the end of August (1989) we lost four ducks and the family cat," Ken Caldicott recalled.

> These were all taken on different days and the cause was traced to wild mink which were trapped and destroyed. On 10 September six of the remaining seven birds disappeared (the seventh was sitting on her nest). These birds disappeared during the day – they were so tame that they put themselves into their shed each night. Any animal attack would have caused the birds to scatter and we would have heard any gunfire. Three days later a lady arrived from the Ministry of Agriculture to take ... samples of duck eggs. Fortunately we had four eggs remaining. (The Caldicotts were subsequently told not to eat the duck eggs.)[70]

The results of the analysis of the grass, foil and eggs disturbed Torfaen council and angered ReChem. On 1 September 1989 ReChem won a temporary injunction which prevented Torfaen council from distributing a report on the PCB contamination. ReChem claimed that the information in the report was incorrect, unlawfully obtained and designed to force the company to close its Pontypool works. Almost two months later, in an out of court settlement, ReChem withdrew the injunction and Torfaen published its report. The council said that the results of the analyses indicated "a most disturbing increase in the contamination levels of PCBs near to the plant" and the evidence was sufficient to warrant a public inquiry. The publication of the results, said the council, "is bound . . . to cause widespread concern to say nothing of the anguish and possible fear it will generate in those persons who have regularly consumed such locally produced foodstuffs as the duck eggs". The ingestion of one of these eggs, the council had been told, would contribute 37 micrograms of PCB to the dietry intake, "a very high figure".[71]

A few days after a further analysis, commissioned and paid for by PEPA, showed high levels of PCB contamination on the Caldicott's land, ReChem said they would be delighted to co-operate in a special investigation by Welsh MPs into their operations. A House of Commons all-party select committee on Welsh Affairs had decided to go ahead with a special inquiry into the plant because of a new controversy. On 13 December 1989 the committee questioned ReChem executives and officials from Torfaen council while MACATW demonstrated outside ReChem's gates. ReChem had successfully defended their activities and refuted accusations that they were responsible for contamination locally, but the contamination of their liquid effluent and of their own site was a different matter. David Powell:

> In 1988 internal memos were leaked from ReChem – not known by whom, but they showed concern by a ReChem scientist about the levels of PCB and dioxin contamination at the site, apparently shown in some test readings. They also showed that Malcolm Lee wasn't concerned and in fact he said that he had been assured that there wasn't a problem over site contamination. ReChem, when asked by television news at the time, said there had been an arithmetical error in the interpretation of test results. Two days later they were forced to admit that there was contamination but it wasn't of concern as it was only a blip.
>
> As far as liquid contamination is concerned there are two pieces of information which suggest that ReChem's discharges through sewers have at times been above the consent levels. The first of those was, again, a leaked piece of information and the second was extracted by the Welsh Affairs committee.[72]

The evidence, it appeared, had mounted up against ReChem yet the company stuck to its guns and argued that the PCB contamination was not their doing. They were able to dispute the findings of the city analyst in Cardiff by placing their own set of results beside his.

When the Welsh Office reported that analysis of duck eggs, collected from the Caldicott's new ducks before Christmas, showed PCB levels of 380 parts per billion (the earlier tests had shown 280 ppb), Shirley Caldicott told the local press that the ducks seemed to be laying eggs without shells.[73]

Despite the evidence that PCB contamination in the Panteg area was apparently higher than normal, the Welsh Office did not believe there was enough evidence to warrant a public enquiry. On 23 April 1990 MACATW led a protest group of 30 people outside ReChem's gates. The group said it would keep up the demonstrations until the Welsh Office agreed to a public enquiry.[74] The following month Torfaen council released new figures which showed the highest PCB contamination levels ever recorded in the area around ReChem. Tests taken taken at the same time from similar areas by ReChem, however, showed lower levels.[75] The council and ReChem agreed to mix up their samples and split them three ways, the third set to be retained for independent analysis.[76]

At the end of May, ReChem announced they had entered a deal with an Italian waste disposal company to build an incinerator in Italy. Two days before the Welsh Affairs committee report was due to be printed Torfaen council and ReChem still disagreed over the levels of contamination. ReChem claimed their results had been vindicated by the government laboratory at Harwell. The council argued they hadn't. The Welsh Affairs committee decided it was time for the government to settle the issue. The committee recommended that independent experts undertake a comprehensive monitoring programme of the plant and its environs. "The independent experts should produce a report showing the results of the monitoring programe in a form which is publicly available and readily comprehensible to the public. The report should be presented to the Secretary of State who should order a public inquiry if the results show that there is a serious risk ro public health or the environment."[77] It was no surprise to the community that the committee also recommended that "major incinerators are not in future located near residential areas".[78] For some protestors this was enough to convince them that ReChem should close.

After the summer, the debate intensified when the council learned that levels of PCB contamination were rising. ReChem, the councillors argued, should stop burning PCBs until the independent monitoring had been carried out. STEAM supported the call. ReChem said there was no reason to suspend operations.[79] But two months later, when Torfaen council produced a fresh batch of statistics, they showed that despite the joint sampling strategy there were still discrepancies between the council's results and ReChem's results. The council's results showed PCB levels in duck eggs to be 480 ppb and 210 ppb while ReChem's showed 9 ppb and 6.5 ppb.[80]

Notes and references

1. This account is based on interviews and correspondence with David Powell in 1990 and 1991.
2. Ibid.
3. Ibid.
4. *Sunday Express*, 5 August 1984.
5. Ibid.
6. Ministry of Agriculture, Fisheries and Food Veterinary Investigation Service, Gloucester, V.I. Centre reference no 311083.
7. *Sunday Express*, 5 August 1984.
8. Welsh Office, letter to Colin Haines, 29 March 1984.
9. Welsh Office, letter to Colin Haines, 11 April 1984.
10. Welsh Office, letter to Mark Robinson, MP, 20 July 1984.
11. Welsh Water (Dwr Cymru), letter to Colin Haines, 9 August 1984.
12. Welsh Office statement. See *South Wales Argus*, 20 August 1984.
13. Welsh Office, Agricultural Department, letter to Robert Stephenson, 11 September 1984.
14. *Western Mail*, 26 September 1984.
15. Friends of the Earth report on pollution in Pontypool, Brian Price, September 1984. FoE, 26-28 Underwood St, London N1 7JQ.
16. See *Western Mail*, 29 September 1984.
17. Ibid.
18. See "Dioxin village sounds the alarm on deformities", *Sunday Times*, 23 December 1984 and "The Cyclops Children", *New Society*, 17 January 1985 pp 104-5.
19. *Sunday Times*, 23 December 1984.
20. Jones, Anthony, letter to Nicholas Edwards, Welsh Office, 24 October 1984
21. *South Wales Argus*, 24 December 1984.
22. *South Wales Argus*, 29 December 1984.
23. *Western Mail*, 14 January 1985.
24. Ibid.
25. Presentation to the Secretary of State for Wales on the operation of ReChem International Limited and the need for a public inquiry,

Torfaen Borough Council, pp.1-2.

26. Powell, David, open letter to the people of Canada. He sent subsequent similar letters to the Belgian and Swedish people.

27. On 5 December 1985 the ReChem management team of Richard Biffa and Malcolm Lee bought the company from Reclamation and Disposal Ltd, a subsidiary of British Electric Traction (BET), in a deal believed to be worth £1.8 million. ReChem Environmental Services, Biffa and Lee's new company, took on PLC status on 18 May 1988 with the placing of 5,661,600 shares (21.3 per cent of the issed share capital of the company). In 1988 ReChem was valued at £51.8 million. In 1990 ReChem recorded record profits of £5.7 million on a turnover of £21 million. When ReChem merged with Shanks and McEwan, the Scottish waste management company, in December 1990 the deal was worth £172 million. ReChem now operates as a separate division within the enlarged Shanks and McEwan group.

28. See note 26.

29. Powell, David, interview and correspondence with Robert Allen, 1990 and 1991.

30. Ibid.

31. Powell, David, open letter to Canada.

32. See Chapter 10.

33. ReChem International Limited; Response to Greenpeace document, p.8.

34. Ibid.

35. Ibid, p.11.

36. Ibid.

37. Hazardous Waste Inspectorate. Hazardous Waste Management: An Overview, 1st report of the HWI (1985), p.34.

38. Phibbs, S; The public and the polluter: a local perspective on ReChem's plant at Pontypool, working paper 467, School of Geography, University of Leeds.

39. HWI (1985).

40. Link, Ann, correspondence with Robert Allen, 1991.

41. Powell, David, interview with Robert Allen, 1990.

42. See *South Wales Argus*, 26 November 1984, *Free Press of Monmouthshire*, 30 November 1984.

43. *South Wales Argus*, 24 November 1984.

44. *South Wales Argus*, 6 December 1984.

45. *South Wales Argus*, 26 November 1984, *Weekly Argus* 29 November 1984.

46. Second Toxic Watch on ReChem, STEAM, 8 June 1985.

47. Ibid.

48. Conference on the incineration of toxic wastes, County Hall, Cwmbran, 27 April 1985.

49. Powell, David, interview with Robert Allen, 1990.

50. Welsh Affairs Committee: Second Report: ReChem, 6 June 1990,

p.16.
51. Garland, Judge: Judgement Torfaen Borough Council v. ReChem.
52. Torfaen/ReChem pp.9-14. See also Welsh/ReChem, pp.48-50.
53. Ibid.
54. Ibid.
55. ReChem, letter to Torfaen Borough Council, 23 October. 1989. See also Welsh/ReChem, pp.53-54.
56. Welsh/ReChem, p.20.
57. Ibid.
58. Welsh/ReChem, p.21.
59. Ibid.
60. Welsh/ReChem, pp. 22-23.
61. Powell, David, interview with Robert Allen, 1990.
62. Welsh/ReChem, pp.13-14.
63. Powell, David, interview with Robert Allen, 1990.
64. Welsh Office, letter to David Powell, 16 November 1988.
65. Department of the Environment, "Statement on the movement and disposal of toxic wastes", 14 August 1989.
66. Minister of the Environment (Canada), letter to Llewellyn Smith, MEP, 27 November 1989.
67. Powell, David, letter to George Mason, *Hazards Bulletin*, Workers Educational Association.
68. Belgian Ambassador, letter to Sarah Preece, 14 August 1990.
69. Hunt, David, Secretary of State for Wales, "Panteg Environmental Investigation" Welsh Office W9120, 30 January 1991.
70. Diary and documentation compiled by Ken and Shirley Caldicott.
71. Torfaen Borough Council report to Secretary of State, pp. 5-8.
72. Powell, David, interview with Robert Allen 1990.
73. *South Wales Argus*, 6 February 1990.
74. *South Wales Argus*, 24 April 1990.
75. *South Wales Argus*, 9 May 1990.
76. *South Wales Argus*, 10 May 1990.
77. Welsh Affairs Committee, ReChem report, p. xi.
78. Ibid.
79. *South Wales Argus*, 21 September 1990.
80. *South Wales Argus*, 7 November 1990.

Chapter 4

ELLESMERE PORT
and CLEANAWAY
Twin Peaks: The story of a town
with two incinerators

Looking directly across the river Mersey from Liverpool's Speke airport you see Ellesmere Port. In the north-west conurbation, Ellesmere Port is a microcosm of modern industrialized activity. It has been a place of commerce and industry for several centuries; the Manchester ship canal effectively begins at Ellesmere Port. On the southern bank of the canal is a horseshoe shaped site where the Redland Purle/Cleanaway waste disposal company have operated a high temperature incinerator since 1974.[1] There is no incongruity about the Cleanaway operation. In an area of towering, smoking chimney stacks, giant heavy duty pipes and rusty railtracks, smoke, smells, dirty water and frothing canals are to the people who live and work in Ellesmere Port nothing out of the ordinary. Looking north and north-west from the M53 motorway which snakes out of Ellesmere Port past Calor Gas, Associated Octel and Cleanaway, dissecting the multi-tracked railway and the Shropshire Union canal, the vista is of heavy industry all the way to the junction with the M56 to Manchester. The massive ICI works at Runcorn dominates the skyline. Heavy industry on Merseyside is a way of life; pollution is a byproduct of existence. There is also a thin line between the pay cheque and the dole cheque.[2]

Redland Purle came to Ellesmere Port in 1972, when planning permission was given for a chemical incineration works. In December 1981 it was renewed for an indefinite period. A year later the company proposed a drum storage and handling compound to Cheshire County Council; as the majority of the liquid waste delivered to Redland Purle was brought by bulk tankers, Cleanaway argued that they needed to offer a service to potential clients who produced small quantities of waste that could only be delivered in drums. The proposed compound would hold up to 500 45 gallon drums. Chemicals stored would include chlorinated hydrocarbons, alcohols, esters, cresoles, phenols, organic acids, ethers and ketones. Any fumes from the handling of these drums would be directed into the incinerator with a system designed in accordance with guidelines set up by the Health and Safety Executive (HSE). The Cheshire County planning officer consulted the appropriate regulatory bodies, neighbouring companies and the local water authority. No one had any objections.

Ralph Ryder had worked at Calor Gas on a site adjacent the Redland Purle/Cleanaway works for seven years before the incinerator started.

This incinerator, we were told, was to deal with liquid waste from the local pharmaceutical industry. I worked as a plant operator. My duties were wide ranging, from the filling and testing of gas bottles, painting and reconditioning them to the offloading of empties and loading the full ones. This chemical waste disposal plant hadn't been working long when we, the loaders, found that on dull overcast or humid days, if the prevailing wind was in the right direction, the steam plume from the incinerator would swirl down onto the loading bays where we worked. Whenever this happened, which was frequently, we found ourselves unable to breathe; we would suffer from a severe burning of the nose and throat, our faces would tingle and our eyes would smart. There was a terrible feeling of tightness on our chests and lungs, followed by the most horrendous headache and aching joints later in the day. Nausea overtook some of us if we got caught in the swirling cloud of steam for more than a few moments. This resulted in a horrible sickly taste in your mouth, and vomiting occurred on a number of occasions. We complained to our management that it was impossible to work in these conditions; this plume was making us very ill. Our job, loading of bottles, was done by hand and was a very physical job; each bottle weighed on average 70 lbs and throwing them three feet high meant we were huffing and puffing quite a bit throughout the working day. Management accused us of "swinging the lead" when we reported that the plume was affecting us. "It's only steam," the plant safety officer would say. "Suck a lozenge and get back to work or up the road," was the foreman's answer. He would walk down to the loading bay, spend two minutes there, then report to management that there was "no problem with the plume". The night shift, which was made up of temporary labour, was forced to carry on working in the plume even though it was making some of them very poorly. These men were amazed when they later came on days to find that the regular workforce refused to work in the plume. Over the following months it became impossible to work in the dreaded plume so all the loaders rallied together and walked off the job, retired to the canteen; there we closed all the windows and sat tight. This happened many times due to the unbearable conditions caused by the plume. We were often threatened with the sack "unless you go back to work". It took a lot of courage to refuse as we were mostly young married men with children to support. And we were so afraid we would permanently damage our health trying to work in these conditions.[3]

The incinerator had disturbed the people of Ellesmere Port for over a decade when Cleanaway announced in the early eighties that it wanted to build a new incinerator on another site. "We have enough hazards, smoke and muck in the Port without any more," was the view of councillor Isobel Reeves, an opinion shared by the Calor workers who, along with the nearby residents, had made the most vociferous complaints about Cleanaway's operation. Ralph Ryder:

We got in touch with the local council with the naive idea of getting it shut down. After many months and many complaints they agreed to meet our representatives and then they visited the incinerator site, as did Calor management. After the visit the councillors and Calor said: "The plant is wonderful, its working okay, there was no real reason why we shouldn't work in the plume". It was, they said, "only steam after all". This was totally unacceptable to us and although our complaints continued to fall on deaf ears, we still refused to work in the plume.[4]

In June 1977 Calor Gas, who were losing money because of lost production, brought in a consultant to test the contents of the plume, and to find out if there was any danger to the workers' health. "Unfortunately," recalled Ryder, "on the day of the survey the plume did not descend low enough to enable any atmospheric test to be made, . . . so we carried on with our action."[5] It worried the workers that Calor might retire them on medical grounds without adequate redundancy if their health did suffer from the emissions.

In early 1981 Calor Gas eventually came to an agreement with Redland Purle/Cleanaway. When the plume was blown onto our workplace and hung around for long enough to become unbearable, if we gave them a phone-call they would reduce the output of steam so that the plume didn't descend onto our plant. Of course, the arguments raged as to what was acceptable to us, the workers, and what Calor management considered acceptable. Sometimes when the plume was going straight over the plant at the height of about 80–100 feet, we could see in the sunshine a fine mist falling on and around us. Even on days like this we noticed that in the space of a few hours we would start with the sore throats, headaches, etc. After trying to work we would retire to the canteen on what management deemed "a nice day" with no swirling plume problems.

Things remained pretty much at the *status quo* with us, the plume, Calor management and the council. Then investigations by the workforce at Calor revealed that other industrial workers in the area, notably Shell, Associated Octel, Burmah Castrol and Van Leer, as well as some residents of Westminster Road, Cresent Road and Wellington Close were also complaining bitterly about the emissons from the incinerator and the health problems they were causing. In spite of this, in December 1981, the council saw fit to give planning permission for the site to be used indefinitely for the incineration of poisonous wastes. There was no research or investigations into any of the complaints. We were outraged and complained bitterly to the council. Our spokesmen had meetings with various councillors but all to no avail, all did nothing. "We have visited the plant and everything is working properly, they are burning to government's specifications" they'd say.

When the company was taken over the new people accepted the practice of the phone-call to reduce the plume; unfortunately no record has ever been kept of when or how often we have had to ask for the plume to be reduced. Cleanaway had operated an incinerator in Essex for six years before they purchased the Ellesmere Port site. They closed the Essex site down and I have never understood why they closed a multi-million pound incinerator after only six years. Maybe it was in a Tory area. Anyway, having new owners didn't make any difference to the plume situation with us.[6]

It was a miserable, rainy autumn day in 1983. A heavy plume was swirling above the Calor loading bay. Ralph Ryder, who was loading 29lb cylinders onto a trailer, bent down to grasp the base of the cylinder. A large raindrop which had fallen through the plume went straight into his ear. "It felt as if someone had poured red hot liquid into it," he recalled. That evening he began to suffer with discharges from the ear and from earache. The problem persisted for several days, but he wasn't unduly worried because he had a history of ear complaints. After a few weeks with the pain and the discharges increasing he visited his local GP who told him that the eardrum was badly inflamed and infected. The doctor said he believed the inflammation had been caused by excessive chlorine, probably from the local swimming pool where

Ryder was a frequent visitor. For a year the problem persisted. The infection was diagnosed as chronic. Ryder was told he would have to have an operation on his middle ear. This cured the worst of the infection, but left him with tinnitus (whistling and noises in the ear). In February 1990 Calor terminated his employment on medical grounds.

Back in 1984 one of the Calor workers was looking through a copy of his wife's magazine, *Woman's Own*, in which was a story about an incinerator in Scotland where the local residents were suffering from burning throats, headaches, aching limbs. It sounded familiar. The feature mentioned dioxins and PCBs and the effects these chemicals had on the human body. "Reading this article scared the living daylights out of the workers of Calor Gas," said Ryder.

> We started – we were worried men – to check the wagons going into the incinerator near us and we were horrified to see loads of 45 gallon drums with PCBs in yellow crayon written on the sides entering the plant. What in God's name had we been working next to for all these years? What chemical particulates had we been breathing every time the plume came down on us? Why had those people, who must have known better, been telling us it was "only steam" for all those years?[7]

The Calor Gas workers decided to extend their campaign in an attempt to have the incinerator closed down. A local GP and a local MP, Mike Woodcock, joined the spokesmen for the Calor workers, Billy Wilkinson, Ralph Ryder and Stan Dixon, in a discussion about the problems. Complaints were sent beyond the council to the HMIP who sent an inspector. Walking into the loading bay with the Calor trade union representative, the HMIP man sniffed the plume, exclaimed: "Good God, that's terrible", rushed back to his car and left. The Calor men were convinced something would be done. After all the pollution inspectorate had now seen and smelled the problem for itself. Yet to the workers' knowledge no report of his visit or of the incident was ever published.[8]

"Bloody hell, Stan, did you see the state of that plume then? I've got it on tape," Arthur Collins, the duty security man, announced as the foreman fitter staggered out of a white mist up to the gatehouse. "See it Arthur, I swallowed it," Stan told him after he had been revived by oxygen. The Calor workers, incensed by the incident, attempted to arrange a meeting with the councillors to show them how bad the plume could become and to demand action. They had the evidence on tape. But Calor management said they couldn't have the tape. It was, they said, their property.[9]

The Ellesmere Port and Neston Local Plan is quite precise on the subject of hazardous development. According to the Borough Council's June 1989 draft plan "the character and history of industrial development in the Borough has resulted in an unusually large number of hazardous installations" which "cover large areas of Ellesmere Port including the town centre". The council's role on hazardous development, it states, is "largely restricted to determining applications for planning permission. In considering applications the Council consults the HSE (among others) as appropriate and takes their views into account. This does not mean that the Council is bound to accept HSE recommendations". Under the appropriate legislation this means that "all development proposals which result in an increase in the number of people at risk are referred to the HSE". The HSE in turn carefully consider "the risk to occupiers or users of the new development and nearby population and they may in some cases recommend that the application is refused". The regulations "under which local planning authorities consult the HSE are contained in Department of the Environment circular 9/84 *Planning Controls Over Hazardous Development*". The draft plan is also clear about recent amendments to the planning legislation "to prevent the introduction of hazardous substances onto a site without planning permission". Hazardous installations, say the draft plan, are:

> those industrial plants where part of the process carried out involves the bulk handling or storage of certain defined chemicals – these may be explosive, radioactive or toxic. These chemicals and the products produced from them are fundamental to our modern lifestyle. Nevertheless, the escape into the environment of these materials can cause nuisance, illness, serious injuries or even deaths among the general public.
>
> Despite these special controls, hazardous installations still pose potential risks to the general public beyond the site boundary. The Town & Country Planning function is important in seeking to reduce potential risks by preventing new development in proximity to hazardous installations. It is also important in preventing new hazardous installations being developed in inappropriate locations.[10]

Despite this concern over hazardous installations, which would have included the Cleanaway operation, the council was – in the opinion of many Ellesmere Port residents – ignorant about incineration and its effects on the environment. When Cleanaway submitted an application in January 1988 for outline planning permission for a new incinerator on a nearby site, it was supported by consultant chemical engineers Cremer and Warner. The consultants stated that they considered the information presented in Cleanaway's application to be "both realistic and comprehensive". The environmental control objectives, said the consultants, would be to "minimise" the environmental impact of "gaseous emissions" from the new incinerator; this would be achieved by adopting the (West) German air pollution standards which, said the consultants, "are generally considered to be among the most stringent of those currently in force around the world". Cremer and Warner further stated that the adoption of "such stringent standards" and "satisfactory compliance with both the present UK standards (as reported in Best Practible Means (BPM) 11), as well as with any improved standards which HMIP might reasonably introduce within the foreseeable future should be reassured (provided, of course, that the equipment installed for the purpose proves capable of achieving the levels of performance required)". Cremer and Warner added:

> Accordingly, we believe the information currently provided herein should now be adequate for the local authority to decide, in principle, on the acceptability of the Cleanaway proposals, and to respond positively to their present application. From the authority's own past knowledge of Cremer and Warner's integrity, independence and complete objectivity in dealing with similar applications, we would also hope that our continuing involvement in this project, albeit as technical advisors to Cleanaway, will provide some additional assurance that all the relevant safety and environmental issues will be fully and properly assessed, and that the plant will be designed and built with all necessary controls and countermeasures incorporated to ensure full compliance with the requirements of the various regulatory authorities such as the HSE, HMIP, North-West Water Authority (NWWA), etc.[11]

The Waste Disposal Authority (WDA) for the area, Cheshire County Council, decided they too needed an independent consultant who would furnish answers to questions drawn up by the County Council in consultation with the Borough Council. Of the 12 questions submitted to the consultants, Selectamaster Process Engineering, two were particularly relevant to the local community.

> The site is located in a primarily industrial area with residential properties located approximately 1,000 metres to the west of the site. There is a history of odour and dust complaints from the public which have occurred as a result of accidental emissions from some industrial premises. There is a substantial office block occupied by Shell within 1,000 metres to the east of the site. On the information supplied, are there likely to be any air pollution problems in the residential area to the west of the Shell office block to the east?[12]

The County Council also asked: "On the basis of current information and advice available, is there a possibility of any acute or chronic effects on the public due to the long-term incineration of wastes on the site?"[13] The Selectamaster report, which was prepared by Drs Stan Kolaczkowski and Barry Crittenden in April 1988, was based on the expertise and experience of the authors, replies received by Cheshire County Council from North-West Water, HMIP, Civil Aviation Authority, HSE and the Fire Brigade HQ of the council, the Cremer and Warner report and discussions with Cleanaway and Cremer and Warner, including a visit to Ellesmere Port.

On the first question the consultants stated that there "should be no contribution to the history of odour or dust complaints in the area" and it was "unlikely that the residential area to the west of the proposed site will be affected by air pollution problems". Good dispersion, they added, should also ensure that the Shell office block is unaffected by air pollution problems. The answer to the second question has since been quoted extensively in support of Cleanaway's proposals:

> There is no information or advice currrently available which suggests, without doubt, that there would be a possibility of acute or chronic effects on the public due to long term incineration of wastes in the proposed site. In recent years there has been much speculation by the media and public about risks to the health of animals and public in areas surrounding waste incinerators. However, to the authors' knowledge, no scientific or medical study has indicated that there are special problems with hazardous waste incinerators. It is worth noting that the conceptual design specification for the proposed incinerator should ensure much better destruction and removal of wastes than other incinerators currently operating within the UK.[14]

The problem for the Calor Workers Action Committee, the Ellesmere Port Clean Air Committee (formed by Ralph Ryder), and other local groups was the inability of the councils, the consultants to the councils, Cleanaway and everyone else connected with the new incinerator to include the community in their deliberations, to hear their fears, to listen to their, albeit anecdotal, evidence; it seemed that the people who lived and worked in the area did not have a right to be involved in a process that they were sure would be injurious to their health. The Selectamaster report did not ease their fears. The groups wanted continuous monitoring of dioxin and furan emissions yet the recommended analysis was "at least twice a year". Kolaczkowski and Crittenden admitted, in their answer to emissions and concentrations from the new incinerator, that some deposition of particulates, metals, metal compounds, phosphorus compounds, acids and polyhalogenated organic compounds must be expected. Because the final design specifications had not been completed the consultants said they could not predict typical depositions in the area close to the new incinerator. They did note that it would be extremely difficult to differentiate between the sources of deposition of any particular compound. The authors added that Cleanaway and their consultants Cremer and Warner would seek incineration technology which would meet current proposed operation and emission standards and that the design would include a rotary kiln, an after burner, a heat recovery unit, scrubbing units and a stack. A rotary kiln, the authors said, was a proven method of destroying hazardous waste in Europe and the USA. In essence, the consultants agreed, the proposed incinerator and the self-imposed standards (British, German and US) would "ensure a more complete incineration of wastes". Did this mean, the campaigners wondered, that the old incinerator operated under conditions and standards that allowed it to emit concentrations of toxic gases that were higher than the new limits and did that, in turn, mean that Cleanaway were admitting they had released toxic gases into the Ellesmere Port air?

The following month the Cheshire County Council planner, in an eight page document, recommended that outline planning permission be granted, that the county council enter into an agreement with Cleanaway (under section 52 of the Town and Country Planning Act 1971) to close and demolish the old incinerator within a year of completion of the new incinerator and revoke all planning permissions for the old site. The planner also sought assurance from HMIP that:

> The proposed incinerator will fully meet the standards established in the Inspectorate's latest Best Practicable Means for chemical incinerators, known as BPM 11.
>
> The levels of halogenated species in wastes should not be allowed to exceed the capacity of the plant's scrubbing unit to remove them.
>
> The County Council will have access to any measurements provided by Cleanaway Ltd to the Inspectorate as required by BPM 11 on emissions.
>
> The County Council will have access to any atmospheric sampling provided by Cleanaway Ltd to the Inspectorate if required by BPM 11 to monitor the effects of the plant on surrounding area.[15]

The planner also stated that when the company applied for a site licence (under the Control of Pollution Act 1974) that a hazard and operability study be undertaken, that an operation and procedures manual for normal operation, start-up, shut-down and emergency circumstances be provided, and for there to be measures to ensure that hazardous wastes could not be inadvertently fed into the incinerator during start-up, shut-down or emergency circumstances.

It appeared, therefore, that the County Council had no objections to the location, despite the fact that the new incinerator would be closer to the town, its shopping centre and schools, and no objections to Cleanaway's proposals to build an incinerator capable of taking 50,000 tonnes a year of liquid, sludge and solid wastes, that would replace the old 18,000 ton liquid waste incinerator.

Following consultations with various bodies, the County Council was told that the North-West Water Authority was unable to make detailed comments until they had received detailed information about Cleanaway's proposals for the disposal of effluent from the site; that the HMIP had no objections to the proposal; that the Factory Inspectorate of the HSE were satisfied that the proposed incinerator would meet the requirements of the Health and Safety at Work Act 1974, both for employees at the plant and for persons working or living in the area; that the Civil Aviation Authority had no objections; that the Council's Waste Disposal officer had no objections; that the Council's Chief Environmental Officer saw merit in the replacement of the old incinerator with a "new, improved" incinerator; and that the County Fire Officer had made detailed comments about adequate firefighting facilities.

The only objections the council had received had come from 250 employees of Calor Gas and Associated Octel, from a resident who said the proposal posed an unnecessary risk to the health of local residents from air-borne pollution, and from the Willaston Residents' and Countryside Society who said that emissions from the plant would have an undesirable effect on the environment and that the sludge residues from the incineration process would cause pollution problems when landfilled. A petition from the Burma Castrol workers was lost. According to Ron Carr, a Little Sutton resident who was present at the planning meeting, none of the objections were mentioned. "The whole thing was rubber stamped."

Safe running of the proposed incinerator was the stated desire of the County Council, in its role as a WDA. The planner made this clear when he said that strict adherence to the design of the incinerator and to the operating procedures for the receipt, handling and combustion of the waste materials was necessary. In his observations he added:

The planning system has, however, only a limited role in regulating the environmental impact of the proposed incineration plant. Of much greater importance in minimising the impact of the plant on the environment is the Inspectorate of Pollution, which is responsible for air pollution aspects of the plant, and the County Council as site licensing authority, which is responsible for the day to day operation of the plant. It is, therefore, proposed that the Inspectorate of Pollution be requested to provide assurances to the County Council on air pollution measures and that Cleanaway be requested to provide additional information on operational matters with their site licensing application. These assurances and further information can only be provided once more detailed design work has been done on the plant.[16]

The County Council had followed the letter of the law and had asked for more control over Cleanaway's operation (more than any other WDA in the country over toxic waste disposal operations) but it was not what the community wanted. Ignoring public opinion, the council had given Cleanaway permission to proceed. The absence of a public debate annoyed the community; the local press had been deluged with letters from people who complained of ill-effects they alleged had resulted from the emissions from the existing incinerator.

In a letter to the local paper, Ralph Ryder questioned the statement by Cheshire County Councillor Derek Bateman that there was no danger from Cleanaway's emissions and the burning of PCBs and other hazardous chemicals. "PCBs," Ryder wrote,

> were banned in the USA in 1983 because they were linked not only with cancer, but with foetal abnormalities. PCBs, when incinerated at less than 1100 degrees centigrade create dioxins and dibenzofurans. The full effects of PCBs on humans is still the subject of scientific enquiry and includes possible effects on the heart, liver, blood vessels and reproductive organs. It causes irritation to the eyes and the vapours can create problems to the nose and upper respiratory tract, was well as internal reactions. Object now; let's find out just what Cleanaway burn – just what comes out of their stack that makes the Calor Gas men ill. Call for a public enquiry now."[17]

He was not alone in his belief that Cleanaway should not be allowed to build a new incinerator and that the company should be investigated by the authorities. Ron Carr also wrote to the local press:

> I registered an objection with the Planning Department at Ellesmere Port, which was passed onto the County Planning Department. A

letter in reply from the county planner was received to the effect that
any objections would be reported to the sub-committee for its consideration. However, I attended the meeting of the Planning Committee
at County Hall and no objection was considered. The item (number
13 on the list) was passed at an amazing speed with no intelligible
discussion and a vote that could only be believed if seen. Other
items ... merited ... concern and detail, I could only conclude
that there was a greater degree of knowledge of these matters, as
opposed to the more technical, but critical aspects of environmental
pollution.[18]

"I saw nothing of these proposals in the press and I am sure hundreds
of other people in the Borough were as uninformed as I was," wrote
Christine Lenihan.[19] (In fact a public notice, under the Town and
Country Planning Act, 1971, had appeared in the local press on 18
January 1988.) R. Smith asked, in another letter, if the community
was aware "that the local authorities have, without consulting the
ratepayers, given their blessing for this obnoxious chemical plant to be
built, and have invited other nations to deliver their waste via Ellesmere
Port docks for incineration within this town"?[20] The chairman of the
Wolverham Residents' Assocation, after a meeting attended by approximately forty people who opposed Cleanaway's plans, wrote that there
was great concern about the new incinerator. "In spite of assurances
by County councillors, the residents of Wolverham feel we already
have enough filth and pollution in this area," he said, and added
that another meeting had been organized: "We would like to invite
councillor Bateman along to answer questions and perhaps he will put
peoples' minds at rest. We would also welcome a representative from
the company as there has not been any discussion with local people
about the building of this plant. We are not trouble makers, as has been
implied recently, but concerned residents who feel we have enough to
put up with already."[21]

In an editorial in the *Ellesmere Port Pioneer*, the leader writer said there
was real concern in the town about Cleanaway's proposals and "the
attitude of Cheshire County Council who, it would appear, agreed with
little or no debate for the scheme to go ahead".[22] Councillor Bateman
wrote to the paper to "dispel anxieties over the possibility of health
risks from the new incineration plant"[23] and Cleanaway's director of
technical services, D.W. Benjafield, wrote from the company's Essex
offices to stress that "we are disappointed that incorrect, misleading
statements have been made by people who simply couldn't be bothered
to find out the facts. This irresponsible, almost hysterical approach can
only generate wholly unnecessary fears and concerns". The concern

among the Ellesmere Port community, he said, appeared "to be based on bad experiences of incineration plants not operated by Cleanaway, in other parts of the country". When the application for the new incinerator was lodged he said "it was encouraging that at least the people who actually have to make the decisions – the County and Borough councils – came to visit the site and see what was actually going on. This is more than can be said for the objectors". Cleanaway, he added "always conducted its operations in full accordance with the highest standards, which are checked regularly by external audit. We have always had an open door policy and do our best to inform people about what we do and what our proposals are for the future".[24] Ralph Ryder replied immediately, quoting the sources of his information and suggesting to Benjafield that he "inform these people" if he believed "their facts are incorrect and misleading; I am sure they'll appreciate advice from such a knowledgeable person".[25]

The public meeting which had been called by the Wolverham group to which the deputy leader of Cheshire County Council, Derek Bateman, had been invited, attracted residents from the community and workers from Shell, Calor, Octel and Van Leer. The workers told of sickness, headaches, sore throats and running eyes which, they alleged, were caused by industrial air pollution. Complaints about the emissions and the subsequent ill-effects were heard from people from all over Ellesmere Port. One man told how a complete shift of workers at Shell reported sick one night as a result of the plume. He held up the accident book to indicate he had the proof. Councillor Bateman told the audience that the new incinerator was safer than the old one. "You either accept the new proposals or you are stuck with the old plant, which you will not be able to close down. I am a local resident too and am just as concerned, but the county council has done everything possible and received all the available specialist advice to ensure that the new plant, when operational, will be as safe as possible". He added that most of the emissions would go over Mickle Trafford, not Ellesmere Port. As the questions from the floor intensified he said he could not answer the specialist questions because he was purely an elected representative and not a waste disposal expert, but that he could arrange a meeting with the experts. "That's what we want" was the cry from the floor.[26]

A public meeting was called by Cheshire council on 16 September at Stanney Lane comprehensive school in Ellesmere Port. The Calor workers had a collection to start a fighting fund. Ralph Ryder:

I was elected co-ordinator of the Workers Action Committee to fight the proposal and I arranged for 4,000 leaflets to be printed and distributed informing the townsfolk of the proposed incinerator and the date of the meeting. I also had it announced on Radio City, Radio Merseyside and the North Wales radio Marcher Sound. The council printed the usual small public notice in the local press. There was some confusion as to when the meeting was due to start so I informed the local radio stations of this and they asked repeatedly for someone in authority to get in touch with them and they would announce the correct time over the air; no one got in touch.

When we arrived at the school at 7.30pm we found that councillor Bateman had started the meeting at 7.00pm with only about a dozen people present. At 7.30pm there was over 400 people and it was standing room only in the hall. The panel of experts included council environmental officers and Henry Pullen, Cleanaway's technical director.

There were no microphones and as all the experts spoke so very softly only the people in the first few rows could hear them. Tempers began to rise when the panel seemed to deliberately avoid the question: "What are the contingency plans for the evacuation of people should anything happen at the plant?" Three times this was asked, three times it was avoided. I made a point of asking this as I knew were no contingency plans because the incinerator wasn't considered a hazard by the local authorities. I wanted to see the reaction if this fact was made public at the school. . . . The panel and councillor Bateman were left in no doubt of the townsfolks feeling towards the proposal.[27]

Councillors Steve Early and Reg Santro told the audience they would vote against the application for detailed planning consent. Councillors Bateman and Tony Sherlock, chairman of the council's environmental health committee, warned the residents that, if the application was refused by the planners, Cleanaway would appeal to the Secretary of State for the Environment who would see no reason for not granting permission. Councillor Bateman added that it was better for the new incinerator to go ahead with all the planned safeguards. A petition signed by 2,410 residents opposed to Cleanaway's plans was presented to councillor Bateman.

In the week before the County Council met to consider the application the Borough Council's controlling Labour group was divided on how to assuage public concern over the proposed development. Several Labour councillors wanted to ask the County Council to defer the decision until an investigation had been completed by independent chemical consultants. Although the company and the County Council had already engaged independent consultants, Labour councillor Brian

Jones wanted a third opinion. The committee voted 11-6 in favour of these extra safeguards.

As the decision-day drew closer Cleanaway attempted to answer the community's questions by opening the plant to the public. Carol Ryder recalled the events of the day:

> Henry Pullen picked up a jar of what he said were PCBs and proceeded to stir the liquid inside with his finger, "There you are," he said, "that's the dreaded PCBs." He offered the jar to the visitors to do the same but they all declined. Another member of staff remarked: "A man could sit on top of that chimney stack all day and it wouldn't hurt him".[28]

Cleanaway had also prepared a 16 page document, presented in a question-answer format, on the existing incinerator, the new incinerator, on Cleanaway's history and operation and on PCBS. Among the questions on the existing incinerator the company answered criticisms that it had been built to lower standards and was therefore unsafe. "The existing performance does conform with current and proposed UK legislation, the current US EPA legislation and most European legislation."

> Is the site safe with all the toxic waste around?
> "The excellent medical history of the employees at the site confirms that there is absolutely no risk even to those persons in close contact with the site's operations."
> Is the steam plume dangerous?
> "The emissions from the existing plant are continuously and automatically monitored for total hydrocarbons and oxygen, which provides a record of the plant's performance."
> What are the levels of importation to the plant?
> "During 1987 the plant accepted less than 200 tonnes of waste materials imported by Cleanaway from outside the UK."[29]

The old incinerator, it appeared, would have a longer life than the council believed. It would not be demolished within a year of completion of the new incinerator but, according to Cleanaway, within 18 months.[30] "I find it incredible," Ralph Ryder said, "that the authorities – both councils, HMIP and HSE – can allow a toxic waste incinerator with a recommended lifespan of 10 years to still be in full operation eight years past its sell-by date."[31]

The communities and particularly the campaigning groups believed they had been presented with a *fait accompli*, by both councils and by Cleanaway. It seemed that they could only meet the residents on their terms and only when it suited them. Neither could the councils or the company answer the direct questions which the community believed they had a right to know. This irked several residents, particularly Ralph Ryder, who noted that no public meeting had been called before outline planning permission had been granted. "Public meetings were held for the building of the M53, the new local supermarket and the new railway station at Overpool. Why not for the Cleanaway incinerator?"[32]

On 28 September Cheshire Council's planning sub-committee took less than half-an-hour to grant detailed approval, subject to conditions, for the new Cleanaway incinerator, which the company said would cost £16 million. Approximately eighty objectors had travelled to the County Hall to voice their opinions. The last option, Ralph Ryder announced after the decision, was to "try and stop Cleanaway getting a site licence". He wasn't hopeful. During his speech to the sub-committee councillor Bateman described the new Cleanaway incinerator as "the Jewel in the Crown" of the toxic waste industry. "We should be proud to have this company in our town. It's an environmental improvement."[33]

Throughout October letters flew around the borough and the county as the campaigners challenged the mandates held by the elected representatives. In the midst of these exchanges Cleanaway reiterated its offer to the Calor Gas workers to visit the plant. Following a meeting of the Borough Council, it was decided that a further public meeting should be held before Cleanaway's waste licence was renewed. A row now broke out between the planning chairman, John Gruffydd, and the other company's opponents. "Although we have, among other things, a condition aimed at preventing 'spurious emissions' from the proposed incinerator, nothing we do seems to be able to prevent spurious emissions from those outside of this chamber and in some cases outside this borough, who seem determined to misrepresent the actions and motives of this authority on the issue."[34]

Letters now appeared in all the regional papers, in Ellesmere Port, in Liverpool and in Chester, as the debate raged. Following a story in the *Liverpool Daily Post*[35] which claimed that the environmental issues had been obscured by local political interests, councillor Bateman wrote a series of letters to local residents and to the press. To one he wrote that people with genuine "heartfelt concerns about this incinerator" have "not been allowed adequate opportunities in a cool, calm and collected atmosphere to ask questions and get responses". The 16 September meeting had, he claimed, been "reduced to a shambles by a small group of people who have been identified locally as members of the Workers' Revolutionary Party".[36] To the *Liverpool Daily Post*, in response to a letter from Ralph Ryder, he wrote that the "genuinely held fears and concerns of local residents" had been "hi-jacked". "I am not criticising Mr Ryder for that; all political parties, including the Workers' Revolutionary Party, attempt to hi-jack issues to put across their point of view."[37]

There was also a contention over the petitions. After the County Council meeting which passed Cleanaway's application, one councillor questioned the authenticity of signatures on the petitions.[38] The campaigners, alleged some councillors, were engaged in a fraud.[39]

While this was going on Ralph Ryder was writing to the Mickle Trafford Parish Council to tell them the news that according to councillor Bateman they would be the recipients of the alleged pollution from the proposed incinerator. "I would like you to know that this committee is totally opposed to the building of this incinerator; we will all be affected by the pollution it causes and to offer 100 per cent backing in anyway we can, if you wish to fight it." The two page, handwritten, letter was a reprise of local events and a brief summary of toxic accidents. Ryder also criticized the borough and County Councils.

> Cheshire County Council and Ellesmere Port Borough Council have both failed to inform the people of Ellesmere Port and surrounding district just what they are allowing to happen here and we as residents (not the Workers' Revolutionary Party as councillor Bateman wants to believe) are doing our best to inform the people and make them aware of the very real dangers that a plant this size will put them in."[40]

If Ryder got no satisfaction from councillor Bateman or from the Borough and County Councils he hoped for more from the local Ombudsman. It was not to be. In his reply the Ombudsman, F.G. Laws, told Ryder that he could not pursue his complaint against the two councils "because I see no evidence that you have suffered injustice through maladministration by the councils". Laws said Ryder's primary complaint that the councils should have arranged a public meeting prior to the granting of planning permission was unfounded. "There is no legal obligation on local authorities to hold meetings to consult the public on planning applications and I am satisifed that the applicants fulfilled their statutory obligations with regard to publicising the proposed development." Laws added that Ryder's allegation that the councils had failed to take account of local residents' and workers' interests before granting planning permission was not true. "The councils also appointed independent consultants to assess the environmental implications," he said.[41]

An attempt to involve members of the House of Commons also ended in frustration. MP Mike Woodcock replied to the Ellesmere Port Clean Air Committee to tell them that it was the council's responsibility "to weigh all the pros and cons before arriving at a decision. Members of Parliament have no role in that process". In what seemed to be a genuine response to the plight of the campaigners Woodcock said that if anything emerged which represented a threat to the people of Ellesmere Port he would "actively campaign against the proposed new incinerator".[42]

The campaigners, it appeared, were powerless and could do nothing about the Borough and County Councils' attitude. It seemed to Ralph Ryder that the apathy in the council chambers had, after the early enthusiasm, spread into the community. Cleanaway would get their new incinerator; the fear that there was a potential danger to health from toxic emissions had been obscured by people whose lives were governed by the twin peaks of power and money. The people opposed to Cleanaway were bound to a system they believed had been put in place to hear their complaints and protect their interests, but in fact was working against them. The bureaucracy and the bureaucrats believed in the stability of their rigid system: there was no flexibility.

A few weeks after they had granted planning permission I asked Cheshire County Council: 'What investigations did you do and what were the findings into claims by workers and residents that emissions from the present incinerator was causing them ill-health?' Their reply was: 'We didn't do any investigations. We left that to the consultants.' 'What investigations did the consultants do?' The answer: 'They didn't do any. They visited the plant and everything was working okay. The men from Calor and Octel haven't been to visit the Cleanaway plant, so Cheshire County Council and their consultants didn't consider their complaints to be of a serious nature.' What is even more amazing and frightening is the fact that the area has never been tested for dioxins, furans, etc. This incinerator cannot possibly be fully efficient as long as no tests are carried out; how will we ever know what dangers the people are in? There has been no environmental assessment because Ellesmere Port already had an incinerator and Cleanaway applied for their licence before the legislation was put in place.[43]

Notes and references

1. In 1972 Ellesmere Port Borough Council agreed an application by Redland Purle for a high temperature incinerator to burn 18-20,000 tonnes of chemical waste. The incinerator was built at the bottom of Dock Yard Road, alongside the canal on the site of an old dry dock, close to Britain's largest petro-chemical complex, Shell Stanlow. Redland Purle had been formed after the merger of the Redland Group and the waste management company Purle Brothers Holdings. In 1981 Redland Purle was bought out 50/50 by GKN plc and Brambles Industries Ltd of Australia and renamed Cleanaway.

2. The north-west of England (Cheshire, Lancashire, Greater Manchester and Merseyside) has the second largest civilian workforce in Britain – 3.1 million in June 1989. The NW also has the fourth highest unemployment rate in Britain and Northern Ireland – 7.9 per cent (January 1990). This compares with 5.4 per cent in Cheshire and 12.6 per cent in Merseyside, which has the highest proportion of long term unemployed (47 per cent) in Britain. The population of the NW is around 6 million, a drop of 1.5 per cent since 1981. Within the NW, Cheshire's population increased by 2.5 per cent between 1981-88 while the population of Merseyside fell 4.9 per cent over the same period. The NW is the most densely populated region of Britain and Ireland with 868 people per square kilometre. It also had the second largest death rate (also behind Northern Ireland). The gross domestic product (GDP) of the NW in 1988 was provisionally estimated at £40.4 billion or 10.4 per cent of the British total. GDP per head was £6,347 or 93 per cent of the British average. The relative position of the NW has gradually weakened over the decade of the eighties which reflects the long-term decline in manufacturing industry's contribution to GDP. In 1980 the NW GDP per head was 98 per cent of the British average.

3. Ryder, Ralph, interview and correspondence with Robert Allen, 1990 and 1991.

4. Ibid.

5. Ibid.

6. Ibid.

7. Ibid.

8. Ibid.

9. Ibid.

10. Ellesmere Port and Neston (draft) Local Plan, June 1989 (Background Papers) 30-31.

11. Cremer and Warner, statement re: Cleanaway application for outline permission for incinerator. See also Cleanaway planning documents, (3/10830), Commerce House, Hunter Street, Chester.

12. See Cleanaway planning documents (3/10830), appendix B.

13. Ibid.

14. Kolaczkowski, S. and Crittenden, B. Selectamaster Limited report.

Responses to questions with respect to the proposed waste incinerator on land off Bridges Road, Ellesmere Port, for Cleanaway. April 1988. See also Cleanaway planning documents (3/10830), Appendix C.

15. Cleanaway planning documents (3/10830).
16. Ibid.
17. "Concern over health of future generations", *Ellesmere Port Pioneer*, 21 July 1988.
18. "No objection was considered", *Pioneer*, 28 July 1988.
19. "Ellesmere Port: A dust-bin?" *Pioneer*, 28 July 1988.
20. "Are residents aware?" *Pioneer*, 28 July 1988.
21. "We have enough filth and pollution", *Pioneer*, 28 July 1988.
22. "Residents are worried", *Pioneer*, 28 July 1988.
23. "Anxieties dispelled on new incinerator", *Pioneer*, 4 August 1988.
24. "Unnecessary fears and concerns", *Pioneer*, 4 August 1988.
25. "Still seeking the facts", *Pioneer*, 11 August 1988.
26. Ryder, R. See also "Protest over incinerator plan", *Pioneer*, 11 August 1988.
27. Ryder, R.
28. Ryder, Carol. Interview with Robert Allen 1991.
29. Cleanaway Limited: Ellesmere Port Incineration Plant (Open Day), 22 September 1988.
30. Ibid.
31. Ryder, R.
32. *We Say No To Becoming A Toxic Dustbin*. Ellesmere Port Clean Air Committee leaflet.
33. Ryder, R. See also *Liverpool Daily Post*, "No surrender vow by waste objectors", 29 September 1988, for an account of the meeting and the protest. Councillor Bateman repeated his comments about Cleanaway's new incinerator in a television interview outside County Hall after the meeting.
34. "Planning chairman attacks opponents", *Pioneer*, 14 October 1988.
35. "Waste issue hijacked", *Liverpool Daily Post*, 30 September 1988.
36. Bateman, D. Letter to B. Aspinall, 18 October 1988.
37. Bateman, D. Letter to *Liverpool Daily Post*, 17 October 1988.
38. See *Daily Post*, 29 September 1988.
39. See note 37.
40. Ryder, R. Letter to Mickle Trafford Parish Council, 20 October 1988.
41. Laws, F.G. Letter to Ralph Ryder, 11 November 1988.
42. Woodcock, M. Letter to Ellesmere Port Clean Air Committee, 14 November 1988.
43. Ryder, R. Interview with Robert Allen.

Chapter 5

BONNYBRIDGE and RECHEM
SCOTTIE, the farmers, the scientists and the workers

"We don't want you, don't come back and if you do come back we'll come with more people and you'll see protests the like of which you've never seen."

Provost, Falkirk

At 3 pm on Monday 17 September 1984 ReChem presented what was, to many of its detractors, an unconditional surrender. After a 10-year war with the community around Bonnybridge and Denny in central Scotland, local politicians and national environmentalists, ReChem had decided to close its incineration process at Roughmute, Larbert in Stirlingshire. The reasons, the company emphasized, were entirely financial and unconnected with "allegations of environmental impact on the locality".[1] This wasn't lost on the community who believed the real war had yet to begin. "On the Monday we heard the news that ReChem was going to close," recalled John Wheeler of the Society for the Control of Troublesome and Toxic Industrial Emissions (SCOTTIE), the Stirlingshire community group. "All through the summer it was really like the build-up to the First World War."[2]

The claim that the pressure from the community had nothing to do with ReChem's decision was one many of the company's detractors found hard to believe, and this was reflected in several media pieces written by reporters familiar with contemporary events in the area around Bonnybridge. The journalists pointed to the company's much publicized plans to expand its PCB processing in Roughmute, announced that summer. ReChem, they claimed, had been forced to make a decision about its future in central Scotland because the workers in Roughmute had refused to handle PCBs until health and safety guarantees had been established; because dioxins had been found in the vicinity of their incinerator; because a local farmer had decided to sue them for alleged damage to his livestock and because of demands for a government enquiry into their operation.[3]

All these facts were true. The workers had held meetings with management about the handling of PCB contaminated waste and had blacked the new PCB processing area; health checks by the company on seven of the 42 workers had found PCB levels within acceptable limits and further tests had been scheduled. Dioxin was present in cattle and soil around the incinerator, farmer Andrew Graham had decided to sue ReChem for £1 million for alleged damages, and there had been repeated calls for a public inquiry, but, ReChem stressed, it had decided to pull out of Scotland because their Roughmute operation had continually failed to return a profit since it had opened early in 1974. The high temperature incinerator in Roughmute was one of the safest in the world, the company insisted.[4]

ReChem's motto "creating a better environment", as far as it critics were concerned, was steeped in hyprocrisy and they were not slow to say so, in the papers, in magazines, on the radio and on television. Throughout the remainder of 1984 and well into 1986 ReChem was criticized for its operations in Stirlingshire and Gwent. ReChem retaliated: criticisms against the company on the BBC programme *Newsnight* in May 1985, and other BBC television and radio programes, were deemed libellous and a writ was served against the BBC, and the company had, during the summer of 1985, won £10,000 in damages from the *Scottish Sunday Mail* for accusations made by a scientist, Dr Larry Robertson. In August ReChem's solicitors wrote to Greenpeace and Friends of the Earth (Scotland). Unless the two environmental organizations withdrew various allegations made on radio, ReChem would sue for libel. When the *New Statesman* reported these actions and sought quotes from ReChem's solicitors the magazine's reporter was told to "get anything you write checked out by ReChem first. It's a very delicate situation at the moment. The best advice I can give you is to keep off the ReChem subject. They're getting very very touchy and they've had enough". It was ReChem's right, a company spokesman added, to protect its reputation against gross defamation.[5] The war, it appeared, was not over.

In November 1985 ReChem responded to a chronology of events concerning its operation and "problems of human and animal health in the Denny and Bonnybridge area of central Scotland, 1970-1983" which had been compiled during July and August 1985 by Jonathan Wills, assistant to MEP Alec Falconer. In a 30-page document ReChem put down their own case "to preserve a true record". Wills' account, they said, lacked impartiality. "The events presented are highly selective, and coupled with annotations and interpolations that are overly prejudicial, serve only to confuse an issue which requires, above all, a dispassionate inspection of all the evidence." Among the crucial points in the document, ReChem stated that as PCBs, dioxins and furans are ubiquitous "their mere presence in the Bonnybridge area is hardly proof of localised pollution. The levels found in soil and in the tissues of animals have to be related to background levels measured some distance away, before a judgement can be made as to whether the Bonnybridge environment is in any way abnormal". The philosophy that a zero level is the only acceptable level, the company argued, is naive and unattainable.

ReChem also stated that the close proximity of a municipal incinerator has been overlooked. "In environmental terms emissions from these two incinerators are indistinguishable." It is known, ReChem stressed, that municipal incinerators emit dioxins and furans, "in fact, 10 to 100 times more than hazardous waste incinerators, despite the generally greater throughput of chlorinated material in the latter. The reason is that refuse incinerators operate at far lower temperatures, and are not designed for the burning of chlorinated wastes". The contribution of the municipal incinerator to localized levels of dioxins and furans had, wrote ReChem, "been consistently ignored".

The dossier, the company noted, had underplayed the enormous work undertaken by the Scottish agricultural colleges in the area. "Several hundred samples of all kinds were analysed for a range of pollutants, including PCBs, dioxins and furans." ReChem acknowledged that dioxins and furans are very persistent in soil, but "the levels reported by doctors Olie and Chittam, the Lenihan report, and others on 1984 samples are so low (in fact, background levels), it is inconceivable that a chronic pollution problem could have existed over the past ten years." Levels in animal tissues, ReChem argued, were also extremely low. ReChem also criticized Wills for his failure to mention work undertaken by Professor Rappe on renal fat from two cows sent to him by Andrew Graham. The company added that there appeared to be a "marked reluctance" among farmers to accept veterinary diagnoses and advice. "The Scottish Agricultural College's report is quite specific about the nature of the problems around Bonnybridge, and discounts the possibility of a single cause. If the dossier had detailed livestock complaints prior to ReChem's arrival in Bonnybridge in 1974 it would have established that the area has had a history of problems."

In reference to Wills' "dangerous occurrences" at Roughmute, the company stressed that its definition of dangerous occurrences was stronger than the legislation. "As used at ReChem a dangerous occurrence was an accident (i.e. unplanned or unpremeditated mishap) which was considered to warrant investigation because of its potential to cause injury to persons and/or damage to plant, equipment or structure whether or not injury or damage has taken place between 1980 and 1984." Of these, said ReChem, only one was notifiable under the legislation.

Other general references in the dossier to ReChem's operations were seen from the company's perspective. "The appearance of black smoke does not mean that toxic materials are being emitted." The "leaked" documents "which appear to have been presented in such a manner as to discredit the company's operational expertise" did not give an objective and constructive assessment of the day-to-day management. A huge area of central Scotland, almost 200 square miles, is claimed to have been affected by ReChem emissions; it is inconceivable, ReChem stated, that its relatively small plant could have had "such a great impact over this vast area". Many of the illnesses described in the dossier were, ReChem wrote, "extremely vague" and "general". The diagnosis of fat cow syndrome in Andrew Graham's herd had been confirmed by several authoritative vets. ReChem also questioned Wills' reference to farms and farmers. "Although the reference numbers go as high as 37, only 21 farms are mentioned in the dossier. Of these 21, 14 have only one or two observations against them and some of these relate to members of the family and not to the livestock. . . . Of the remaining seven farms, the coding suggests that two or three are all under the management of Mr Graham. The absence of detailed locations and names of farmers tends to reduce the value of this report since it is known that Tambowie is nearly 20 miles from Bonnybridge." It is possible, ReChem wrote, to go through nearly every incident described in this dossier and to wonder what the veterinary diagnosis was: "what other factors lay behind the problem described and how many times the farmers in question had any reason to believe other than the fact that their animals were afflicted by the usual range of common spontaneous diseases". ReChem also had a point to make about Andrew Graham:

In Mr Graham's case it seems to be tragic that his reaction to the initial diagnosis of fat cow syndrome, a diagnosis which would be made merely by reading the newspapers, so upset Mr Graham that he seemed to close his mind to the fact that he might have problems on his own farm. Perhaps he did have these doubts, but unfortunately, he does not seem to have initiated the sort of investigation which should have eventually solved the problem. ReChem offered funding to Mr Graham to help get to the bottom of his troubles, and were willing to co-operate with Mr Graham. [In November 1983 ReChem offered £10,000 towards a research programme into the problems with Andrew Graham's livestock.] However that co-operation was refused. It should be remembered that only relatively few of Mr Graham's cows went to West Bankhead and returned to Tambowie to calve down. It appears it was these cows which developed fat cow syndrome. Other cows which never left Tambowie developed other illnesses. It is surely more likely that the cause of such illnesses rested at Tambowie, rather than that they were transmitted 18 miles against a prevailing wind to one single farm to the north of Glasgow, when neighbouring farms had no similar problem. It is noteworthy that cattle which had not left Tambowie, as well as those which grazed at West Bankhead, were affected, although not necessarily with fat cow syndrome. Without expert investigation it seems likely that the truth of Mr Graham's troubles will ever be uncovered.

The remainder of ReChem's response to the dossier detailed the company's analysis of the events; whether Wills' interpretation was, in their eyes, accurate or not, and, in three pages, the problems on Andrew Graham's farm. "This has been made necessary because we have been refused access to Mr Graham's farm and to the laboratory notes and technical reports of the various veterinary experts who have been involved in studying these problems." ReChem concluded their response to Wills' dossier by applauding his initiative. "Unfortunately," ReChem stated, "Dr Wills has exercised such blatant prejudice and selectivity in deciding what he should include – or exclude – as 'evidence', and his personal interpretation of events, that he has negated any of the benefits that such an unbiased, dispassionate narration of the facts may bring."

It is regrettable that while a great deal of effort has been expended by various veterinary bodies, their findings are not available for general public consumption, whereas flawed documents such as the dossier are widely circulated. Therefore, those members of the public not intimately involved with the problem and whose sole source of information is the dossier, have been seriously misled. The purpose of ReChem's annotations to Dr Wills' dossier is to redress the balance, correcting and comenting on statements that would otherwise have gone unchallenged.

The entrenched position adopted by the critics of ReChem has resulted in their refusal to countenance any explanation of the various problems other an unsubstantiated relationship with ReChem operations. This in spite of the overwhelming weight of evidence produced by wide ranging independent bodies of professional scientists. It must be further stressed that the investigations were carried out in the full knowledge that industrial pollution involving PCBs, dioxins and furans could be implicated. Lesions and symptoms typical of poisoning by these compounds have never been observed. All environmental date clearly indicate these compounds are present only at typical background levels.[6]

More than a year after ReChem had closed its Roughmute operation it was able to defend its toxic waste disposal activities. Andy Kerr, FoE Scotland's co-ordinator, had been accused by ReChem of being "grossly irresponsible and malicious" and of making "clearly untrue and defamatory" accusations about ReChem's operation in Roughmute.[7] FoE Scotland protested. "I think most people would regard it as unfair that such a large organization should pick on such a small organization funded by public donations," Andy Kerr argued.[8] It is worth noting that ReChem were not able to intimidate SCOTTIE. "We never made claims we couldn't substantiate, but if we felt we could make claims we would make them. We could substantiate what we said about smoke and fumes."[9]

It is also worth referring to comments made by Sian Phibbs, who, as a research student with Leeds University, compiled a study on the attitudes of the community around ReChem's Pontypool operation. ReChem were not impressed with Sian Phibbs' initiative. "This so-called survey is an indirect attack on ReChem, thinly disguised as pseudo-scientific research," ReChem managing director Malcolm Lee said. "ReChem is under no illusions – the vast majority of local residents would still much prefer we went elsewhere. But we cannot accept without a vigorous response the unwarranted intrusion of blatantly prejudiced and uninformed outsiders, whose 'independently researched' conclusions will be interpreted from what is demonstrably an extremely biased questionnaire."[10]

ReChem was not blind to the attitudes of those who lived around its operations in Bonnybridge, Fawley and Pontypool but it was not prepared to sit back and allow its critics free rein. The company was also not prepared to allow its critics freedom to engage in open dialogue or to interpret the events from their own perspective; anyone who did this was branded prejudiced, biased, selective, ignorant or wrong.

"Cultural theory," Sian Phibbs wrote, "aims to reconcile opposing viewpoints by making people aware of their different (often conflicting) value, moral, political and cultural biases. By doing so, it is intended that people will appreciate that issues such as 'risk' do mean different things to different people, and that there is never a 'right' answer to risk-related problems." Phibbs argued that social survey research had an important role in this process. It could provide material which could mediate between opposing individuals and interest groups. "It is hoped that research on public attitudes and perceptions can enlighten those in authority by revealing the needs and the wishes of local people who rarely have the political power to influence present environmental decisions." From her analysis Phibbs was able to argue,

> that the gap between what the public perceives and wants, and what the authorities believe those perceptions and wants to be, needs to be bridged – and soon – otherwise environmental controversies are doomed to continue on their present course. While social survey research is considered to be an important part of the healing process, the problem remains that resolutions can only be attempted if those in authority will permit the opposing parties to communicate on a more official, more equal and more tolerant basis".[11]

If Phibbs' ideal is seen to be a little naive, given the nature of the toxic waste disposal industry, her conclusions about the role of the academic world is illuminating and provides part of the answer to the problem communities have with waste disposal companies. Academic analysis and debate is not enough, if all each side can achieve is a difference of subjective opinion. As a scientist John Wheeler was disturbed by what he saw as "bad science", the selective methods used by the experts who reported on the ReChem/Bonnybridge allegations. Of an early report, in 1976, he recalled how SCOTTIE had asked the HMIPI if they had tested for heavy metals. They said they had not. "Now that really got me because that was this business of absence of evidence is evidence of absence – that's the problem; if you don't look for it, you don't find it."[12]

The publication of the various reports (including Lenihan's report from the Scottish Office) did not convince everyone that there was no case to answer, that ReChem's incinerator was not a cause of the alleged pollution. In a letter to *Chemistry in Britain* (which became a battleground for opposing views from scientists studying the ReChem issue) Fiona Williams wrote:

> The publication of the Lenihan report has led ReChem to feel exonerated. It is my opnion that this is a false premise – the report has neither exonerated nor incriminated ReChem. There may be no problem with pollution and/ or morbidity levels in the Bonnybridge, Denny area but until further more detailed, scrutiny of the area is undertaken, the problems encountered by the parents of children with microphthalmia and the animals belonging to messrs Taylor and Graham will remain an enigma.[14]

It was a view supported by many in the community, by several scientists and epidemiologists and by the environmental movement.

In August 1985, six months after the Lenihan report had been published and seven months after a study by the government laboratories in Harwell had given ReChem "a clean bill of health", FoE Scotland drew up a list of objectives on the issue of toxic waste disposal:

1. A review of the toxicity of PCBs, dioxins and furans;
2. A strategy for sampling and analysis of PCBs, dioxins and furans, to establish a data base;
3. A comprehensive investigation of the (human) health problems around waste disposal plants;
4. A comprehensive investigation of animal health problems around waste disposal plants;
5. An outline of the criteria which should be met in the disposal of toxic waste.

> Whether the authorities like it or not, there is a lack of trust and confidence in all previous reports on the Bonnybridge problem. This is a new situation, we have become aware of and it requires real initiative. No one group should investigate each aspect above but acknowledged authorities should be allowed to contribute. Honest disagreement is preferable to convenient neatness.[14]

On 17 January 1985 ReChem released a statement from their Southampton HQ on the findings by Dr Alan Eggleton in his report, "The environmental significance of dioxin, furan and PCB levels measured in the vicinity of ReChem's Roughmute plant" from the Environmental & Medical Sciences Division of AERE (Atomic Energy Research Establishment), Harwell. The company quoted one of the paragraphs from Eggleton's report:

> The source of the trace background levels found is not clear, but evidence of a general production of (dioxins) and (furans) in combustion processes carried out at lower temperatures than were used in ReChem's plant, indicate that the source should be sought in such areas as the combustion of natural wood and the burning of domestic refuse in

municipal incinerators, all examples of the latter so far tested having demonstrated (dioxin) and (furan) production. Other possibilities which need consideration are the use of the herbicide 2,4,5-T (which contain dioxin impurity) and proximity to Roughmute of the Grangemouth refinery. General sources of PCB should be sought in industrialised and urban areas, and particular sources at waste landfill sites.[15]

Managing director Malcolm Lee said: "Harwell is the most respected scientific institution in Britain and even our strongest critics must recognise that this report gives us a complete clean bill of health."[16] ReChem's strongest critics, however, did not agree and even Dr Eggleton criticized the lack of facilities and money for carrying out such analyses in Britain.[17] John Wheeler pointed out that Dr Eggleton had carried out his analysis based on three soil samples, four cow livers and two stack emissions. "In no way can this be acceptable as providing an adequate picture of the concentrations of the various substances throughout the area. . . . There is simply not enough data to allow any assessment to be reliable," he added and in conclusion stated:

> Overall SCOTTIE is very disappointed that Harwell should have seen fit, on the basis of so little data, to have reached such far reaching conclusions. We feel that most people will take these conclusions only as interim speculation. Certainly far from giving ReChem a clean bill of health, the report clearly outlines the need for more data to be obtained, from the environment, animals and humans so that a valid evaluation can be made. This, SCOTTIE stresses, can only come if there is a full, independent, public enquiry and for this SCOTTIE will continue to press.[18]

Among those in the scientific fraternity who disputed the reports from the Glasgow Veterinary College and the Scottish Office (Lenihan) were Dr Alastair Hay, of the Department of Chemical Pathology at the University of Leeds, and Dr Owen Lloyd, unit director of the Environmental Epidemiology and Cancer Centre, Medical School, Dundee. In a paper on "The environment and cancer" Dr Lloyd and his colleagues were optimistic that more comprehensive studies would be undertaken.

> Fearful of the stasticians's frown and dismissive malediction of small numbers, epidemiological analyses of small areas have been underaken only relatively rarely. An example of this omission has been the investigation of the Bonnybridge/Denny situation described in the recent Lenihan report: probably due to shortage of time or to the cramping

terms of reference, the populations investigated in the study contained as many as 40,000 people scattered across wide (80 km squared) tracts of the country, whereas the circumstances called ultimately for small area analyses of geographically-circumscribed communities, perhaps with no more than a few thousand inhabitants. Presumably the appropriate, more detailed, investigations are now at hand.[19]

Dr Hay was critical of the Glasgow report:

. . . The vet said that what had killed Andrew Graham's cows was "fat cow syndrome"; this is basically caused by the animals being over-wintered on a ration that has got too much of one nutrient or another – usually too much carbohydrate or too much protein. The animals metabolise it in such a way that the liver becomes fatty, they become generally unhealthy – you get this fatty sheen to the animals. It usually only affects cows and not the rest of the herd.

Andrew Graham's animals became ill not when they were over-wintered but two or three months after they had been taken out of their stalls and put on the pasture. In contrast to "fat cow syndrome" which is reversible his animals did not get better, they got worse. So we (myself and some colleagues, epidemiologists and a vet) put together a report and we said we thought that the case was non proven, certainly not "fat cow syndrome". We sent our paper to the one person who initially described "fat cow syndrome" and he agreed with us; he said the Glasgow vets were wrong. We sent the paper to the *British Veterinary Record* – which is really the established journal in the UK – and they refused to publish it. They sat on it for a long time, asked us to re-jig it – which we did, and then they said they couldn't publish it. We knew it was explosive, so we sent it to an American journal (*Veterinary and Human Toxicology*).

There was another problem with another farmer who lives right next door to Andrew Graham's land. His animals were dying and the Glasgow Veterinary School said ragworth poisoning has caused their deaths; they based this diagnosis on the appearance of the liver cells. Ragworth produces a number of toxins but the upshot of it is, when you look at the liver cells, instead of having discrete cells together you quite often get cells which have become aggregated and you have one large cell with a number of nuclei in it. So these multi-nucleated cells are one pathological symptom of ragworth poisoning.

This is the sort of evidence that the Glasgow Veterinary School used. They said this is consistent with ragworth poisoning, which it is. However, if you look through the literature – we knew the literature – you can find 25 or 20 chemicals [that would produce the same results], quite a few of which would be produced in an incinerator or found in a chemical dump.

The farmer said there was no ragworth on his land and farmers recognise ragworth because they know it kills their animals; he'd never seen any. There wasn't a major problem with ragworth in the country that year – we checked that. . . . So we said that the ragworth case was not proven. They'd simply used the appearance of liver to come to some sort of diagnosis. The analogy is like knowing my son was wearing a red coat and I'd seen someone walking down the street in Leeds in a red coat and said, "yes that's Tom".[20]

In February 1986 *Chemistry in Britain* published the results of George Smith's and Dr Owen Lloyd's analysis of soil pollution around Bonnybridge, with reference to the case of John Taylor (the farmer who was told his cattle had ragworth poisoning). The authors reported they had found relatively high values of chromium in the soil and high chromium concentrations in the chemical waste dump (which had been used by ReChem between 1974 and 1982). They added their own voices to the strong criticisms made in the Lenihan report of the operation of chemical waste disposal at the dump.

> Our findings of high concentrations of at least one toxic chemical there – all too readily accessible for country ramblers, children and domestic and farm animals – justify those criticisms. Furthermore, large pieces of waste containing those metals had been allowed to encroach on the border of a pasture. The high values of chromium in soils of that field indicated that parts of the grazing land had become contaminated. . . . It cannot necessarily be concluded from these results that the polluted soils (or surface water/vegetation) contributed to the death of the cattle. More veterinary and toxicological studies are required for a satisfactory answer. But this study underlines the importance of using a high density of environmental sampling sites when problems of micro-epidemiology are being investigated."[21]

In what was to become a long running saga in *Chemistry in Britain* and other journals on the alleged industrial pollution around Bonnybridge[22], Smith and Lloyd's paper brought a response from ReChem's consultant Dr Eduljee. He said it was "an unfortunate feature of the Bonnybridge issue that none of the reports detailing the veterinary investigations on Mr Taylor's cattle is generally available" and would the impetus for Smith and Lloyd's study "have been maintained had they seen the actual reports that had convinced the earlier investigation team led by Lenihan of the validity of the ragworth poisoning diagnosis"? Dr Eduljee questioned the hypothesis that metal pollution may have caused the deaths of Taylor's animals. "The analytical presentation of Smith and Lloyd leaves the reader with the impression that metal pollution may have been the cause of the illnesses amongst the cattle during 1977-80, whereas veterinary examination (surely a more reliable indicator) backed up by circumstantial evidence clearly indicated ragworth poisoning."[23]

Smith and Lloyd were joined by three colleagues when they penned their reply to Dr Eduljee's paper. They agreed with the ReChem consultant that some of the reports from veterinary and agricultural colleges on the Bonnybridge problems were not generally available. "Even farmers who provided the experts with free cattle for examination and subsequent autopsy were unable to obtain a copy of the findings on these animals." The authors added that their own enquiries into the relevant veterinary information had indicated "that ragworth poisoning does not have unique signs and that the diagnosis relies partly on the circumstantial evidence that ragworth has been consumed in quantity. Dr Eduljee writes that 'the diagnosis of ragworth poisoning is based on certain specific lesions that are quite unlike those caused by chromium toxicity'. This emphasis on pure chromium toxicity is difficult to justify, as our article did not claim that chromium was necessarily the culprit."[24]

The authors made the point that a photograph of "broad tooth marks across the leaf of one plant", together with the reported presence of the ragworth *Senecio aquaticus* in the field, (referred to by Eduljee) was presented as strong evidence to explain the problems in John Taylor's herd. They argued that this was not consistent with the facts – "cattle almost invariably use the tongue rather than teeth for feeding. What proof is there that the tooth marks were not made by some [other] animal, or were even the track of a caterpillar? To use the leaf of one plant in the context of massive morbidity and mortality is an extreme case of arguing from the very tenuous particular to the general" and that if ragworth did not cause the deterioration of Taylor's cattle, what did? Ragworth is a common weed,

> yet deaths from ragworth posioning are rare. In NE Scotland, where surveys are conducted annually, 18 per cent of the grassland area has ragworth. Not all the species are toxic; indeed *Senecio aquaticus* or marsh ragworth, is not consistently quoted as a toxic hazard; and as its natural habitat is ditches, wet meadows and marshland, Dr Eduljee's claim that it was present in John Taylor's field is incongruous in the context of his argument that the field was liable to abnormal dessication during dry weather. Other ragworth species, such as *S. jacobaea*, are more widely recognised as toxic, but a considerable amount has to be consumed. Dry summer weather during 1978 must have affected the growth of grass on pastureland elsewhere in Britain and yet there is

no suggestion of a general increase in ragworth poisoning in that year. Where is the evidence that the topography of John Taylor's field renders it a site for ragworth poisoning?"[25]

It was the authors' contention that "even if the case for ragworth poisoning at the farm were more persuasive, the wider epidemiological view causes concern". Why, since the late 1970s, they stressed "has the area around Bonnybridge provided such a remarkable cluster of farms with unprecedented problems of animal health – depressed milk production, calf deformities and a loss of over 800 animals?"[26]

This was a question being asked repeatedly by the community in Bonnybridge and Denny and one to which John Wheeler was determined to procure an answer. In March 1985 he had sent a detailed critique of the Lenihan and Harwell reports to the Scottish Secretary of State, George Younger. Like many people in Bonnybridge, John Wheeler expected more from the Lenihan report. It was published on Valentine's day 1985, but to Wheeler's annoyance it was not an investigation into ReChem; "what it was was an investigation into unusual morbidity". When he saw a copy of the report Wheeler was apprehensive. He had agreed, at short notice, to talk about it on the BBC in Glasgow.

"Well," the news broadcaster said to him when he arrived, "we've managed to get a copy." It was three oclock, three hours to the broadcast.

"I remember going [to collect it] and I thought, if this report is really complex I'm in for a hiding here, because they were going to get Lenihan, he was going to be in the Edinburgh [studio]. So I opened it up . . . and I said: 'This is a desk job, it's a desk job'. He's simply asked people for their opinions. But of course once they had Lenihan, they got what they wanted in print. They can still say: "Professor Lenihan said, in his extensive and comprehensive report. . .".[27]

Retrospective criticism of the Lenihan report was widespread throughout the community. The major complaint was summed up by MEP Alec Falconer. "That report, we felt, was insufficient. It didn't carry out the correct epidemiological study. There were questions raised regarding how their statistics were drawn up and how they drew their evidence from the statistics. At the end of the day we were very concerned because there was no in-depth investigation by a body which the community could actually place their trust in."[28]

In 1975 ReChem's Roughmute operation had been up and running less than five months when the first public meeting was called to discuss complaints from the community about "unsatisfactory emissions". It had all happened so quickly. ReChem had looked at six different sites in Scotland before Stirling Council offered them the site in Larbert. The company initially promised 240 jobs, though in discussions with the council 135 jobs were offered. When the company started, they employed 24. The local authority had bought land from the regional health board and sold it to ReChem. Three public notices had appeared in the paper. "They were very misleading," Sandra Farquhar remembered. "If anybody saw them at all they weren't particularly bothered I don't think, but for me it was sudden. One day I was passing the site and I thought, 'what on Earth is this?', it just seemed to appear: 'Toxic Waste Disposal Plant' and the alarm bells immediately started to ring in me."[29] John Wheeler:

> What happened was that ReChem came along and I presume said to the local council, we can provide you with jobs, we've got a plant and we can put it beside your incinerator. The notices went in progression – from a waste processing plant, to a complex for the disposal of chemical waste products, to a toxic and chemical waste disposal plant. Unfortunately that was the old council who were wined and dined by ReChem down in Southampton, and I don't mean anything bad in that, but the councillors were deceived, taken in, naive. The Environmental Health Department, if they got the say so from HMIPI, would have been quite happy with it and I think as soon as the plant got going they realised that they'd got trouble. Sandra [Farquhar] lived just over the back at Bonnybridge from the place and she started stirring it, and so ReChem had this meeting. The local council realised there were going to be problems, and got this liaison committee going, which met twice a year. Eventually there were five of us from SCOTTIE on it and we pulled out. We got fed up. We would ask questions and ReChem would say, we'll bring the answers next time and of course they didn't. These liaison committees are [places] where people go, let off steam; think they're being effective and really are being conned. Unless you really have a grip on the firm, and you have these [meetings] frequently they're a waste of time, a con. But it ran for a few years and we went along with it.[30]

"The reaction of the management of the plant to complaints," recalled Sandra Farquhar,

> was simply to dismiss them and to infer that the complainants could not possibly know anything about the incineration of toxic wastes and that they, ReChem, were the experts. They issued impressive claims about their operations and about their use of "Advanced Technology in

the Control of Hazardous Wastes". This glossy brochure claimed there would be no smell or pollution from the plant, that only steam would be emitted from the chimney stack, that the incinerators were equipped with fail-safe systems and that every possible safeguard was built into the plant design and operational procedure to ensure that the various processes involved did not give rise to any environmental problems whatsoever.[31]

SCOTTIE was formed after that first meeting and two further meetings, which were well attended, followed. On 17 June 1976 Richard Clark, then chair of the group, informed the director of Environmental Health on the local council that "the residents in the Larbert and Bonnybridge area have now formed a society to enquire into the circumstances in which troublesome and toxic industrial pollutants can be liberated into the environment". The society, Clark wrote, would be known as SCOTTIE – Society for Control of Troublesome and Toxic Industrial Emissions. Clark added that the emissions from ReChem sometimes became "unbearable" and were "especially severe on people with chest complaints". He referred the environmental health director to the period between 28 May and 1 June 1976 when part of the village was "obliterated by dense smoke" from ReChem. "This is not an isolated incident as it happens frequently on the Denny/Larbert road and on occasions on the motorway." He concluded: "We wonder how much longer the residents in the area will have to suffer this nuisance and write to ask that steps be taken to ensure that this plant be closed [unless] they can prove what they claim – No smoke, no smell."[32]

The workers at the plant, according to Clark, were "very much anti-SCOTTIE".[33] There was no support from councillors and MPs, and no support from the environmental groups. SCOTTIE was very much on its own, but Sandra Farquhar already knew what she had to do.

Once [ReChem] started operating the plume from the plant just kept coming down to ground level all the time, no matter which direction the wind was going, and particularly in the Spring we got it more. The plant was east of us and we got it all year round, but predominantly in the springtime, when the winds were more easterly it came down on top of us and it really was unacceptable. So I started off by complaining to them as an individual. They used to send up their manager at that time who used to say it's okay, there's no harm in it or it's just teething problems or it's this or it's that. This went on for a while and then I decided that I really had to do something about it, so I started going around collecting signatures for a petition in the areas most affected. A few people came along and helped. That was the first petition; there were others later on.

I made contact with the Lindsays, whose nursery plants were damaged. [In August 1980 the courts ordered ReChem to pay damages of £3,000 to the Lindsays who alleged that emissions in late 1974/early1975 had destroyed a large number of their plants. The company had originally asked them not to publicize the matter and said their claim for damages would be met in full.] And then after about a year, we all got together and that's when we formed SCOTTIE. Before that we were working independently. The Environmental Health Department, very early on, had given me monitoring sheets to fill in, giving wind directions, what the plume was like, if there were smells, and different things like that. I actually did that for year, after year, after year. So there's a record of just about every hour of ReChem, apart from the darkness hours, of almost every day.[34]

Sandra Farquhar's son was about five in late 1974 and was directly affected by the plume. He started to suffer from chest and throat infections on a regular basis and twice was taken to hospital. He had all the symptoms of clinical meningitis, but a lumbar puncture showed he had a very severe throat infection. "Then about six months later, he had very similiar symptoms; I don't actually know whether it was just that the local GP panicked slightly but he was wheeled off again to hospital. Of course at the time I blamed ReChem, but you're never absolutely sure. However, from the day, and I mean the day, that we actually moved from Bonnybridge to here in Dunblane he never had another chest infection."[35]

John Wheeler joined SCOTTIE in December 1976, almost eight months after the group had begun to campaign against ReChem. "I just saw an advert in the *Falkirk Herald* about this society and I went along on purely altruistic motives, I didn't know there was a problem with ReChem."[36] The group had just commissioned an analysis of a hedgerow which showed heavy metal contamination and that night presented a 31-page document. Wheeler was impressed and when asked if he would be the group's science advisor he agreed. In the same document SCOTTIE explained who they were, why they were formed, why they objected to ReChem, what their aims were:

SCOTTIE was formed in May [1976], mainly by people in the neighbouring villages of Bonnybridge, Larbert and Stenhousemuir, where serious atmospheric pollution has caused widespread alarm. The membership of SCOTTIE has grown quickly and now includes interested people from further afield, where similar problems exist. . . .

SCOTTIE believes that the best guarantee of safety for any community exposed to the risk of pollution, is for ordinary people within the community to develop a constructive interest in the problem and in the activities of factories likely to cause pollution. This can be time-consuming and frustrating, as this report will show, but there appears to be no satisfactory alternative to it, because the official watchdogs, such as the District Council Department of Environmental Health and the more powerful and significant H.M. Industrial Pollution Inspectorate, cannot always be relied upon to pursue problems without being repeatedly prodded by the public. . . .

The risk of pollution is an inevitable part of modern life and is the penalty we pay for the benefits we enjoy from our technological society. SCOTTIE excepts this, but is concerned that the risk is adequately evaluated, reduced to acceptable levels and honestly declared. In presenting this first report, SCOTTIE is complying with its constitutional aims, which are to :-

1. Enquire into the circumstances in which troublesome and toxic industrial pollutants can be [released] into the environment.
2. Collect and publish information about the nature and effects of such pollutants.
3. Collaborate with other organisations concerned with environmental pollution.
4. Stimulate research by appropriate individuals and institutions into the hazards associated with toxic waste disposal.
5. Seek to persuade H.M. government, members of Parliament, officials of HMIPI, local councillors and officials of the Environmental Health Department to do all in their power to reduce the level of industrial pollutants to safe and acceptable levels.

In stating these aims SCOTTIE endorses the following view recently expressed in the Fifth Report of the Royal Commission on Environmental Pollution: "It is no longer acceptable that decisions on emissions which directly affect the daily lives of many people should be taken by a small specialist body consulting only with industry; greater participation is needed not least so that the assumptions and problems on which the decisions depend are more widely understood".[37]

SCOTTIE's impressive initiatives were not sustained. The group had given everyone, the company, the local authority, the regulating authorities, a run for their money. By the end of the decade Sandra Farquhar moved out of the area. Within another year the group had pulled out of the liaison committee. ReChem had now been operating for eight years, and the group and its supporters began to despair. A succession of secretaries had come and gone when Wheeler was asked to take it on. Nobody wanted the job. Wheeler remembered: "[SCOTTIE] was just about existing and there was this fellow, a freelance environmental consultant, who said, as we were the only group then doing this kind of thing, it would be a pity to let it die, so I said okay, I'll take on the secretary's job. That was in August 1983 and [by] September 1983 Andrew Graham was playing merry hell in Falkirk, asking down the fire station who was responsible for the ReChem plant. So I got in touch with him and went over to see him."[38] Suddenly there was controversy over ReChem again. Graham impressed Wheeler. A meeting was arranged, Graham's problems were discussed.

Just after Christmas SCOTTIE decided to hold a public meeting. They wrote to Dennis Canavan, the local MP, and asked him to chair it and he agreed if the trade union could attend. They also invited HMIPI, ReChem and Friends of the Earth.

> We had "Today the cattle died, tomorrow?" plastered all over the place. We'd never done anything like this before and I don't think I'd ever met an MP before. The BBC had asked if they could come in.
>
> I remember I went down there and the place was packed. The Provost was there. Because he was a Bonnybridge councillor, he wasn't too chuffed because obviously the council had given the planning permission. The workforce were there. Friends of the Earth were there, Andy Kerr in those days. Les Baker spoke from ReChem; the HMIPI spoke and Malcolm MacDonald, the Environmental Health Director. PCBs were mentioned and Friends of the Earth went on about them. The fact that so many people had come in showed the depth of feeling. The workforce were standing up at the back.
>
> We put the buckets out and I was amazed at the amount of money we got.[39]

It was nine years after the first public meeting in Bonnybridge. SCOTTIE had now been given a mandate by 400 local people including several farmers to call for a full, independent, inquiry. Nine farmers had complained about the mysterious problems with their livestock and had appealed to the Scottish NFU for help. At the public meeting Ian Grant, then vice-president of the Scottish NFU, announced that the union would investigate the problems. The new concern, as far as the comunity, SCOTTIE and John Wheeler were concerned, was PCBs. Wheeler went in search of information about them and compiled a fact sheet. One of these got down to the workforce at ReChem and was put up on the noticeboard at the plant.

It was now the end of March 1984, well over a month after the meeting. The PCB leaflet had been distributed and Wheeler had appealed to George Younger for a public inquiry. He was told by one of the Scottish Secretary's staff that an inquiry would be "inappropriate" but that investigations by the West and East of Scotland agricultural colleges were already underway.[40]

In April SCOTTIE called another meeting, this time in Bonnybridge Parish Church. John Wheeler presented the information he had gathered on PCBs. It was suggested that a petition be put together, which would not be to close ReChem down, but to bring attention to the problems in the area, the PCBs and the fact that they were being brought in from outside.

Concern about the possible effects of industrial pollution in the area intensified when the mother of a deformed baby, born locally, contacted Andrew Graham to raise her suspicions about the cause. Wheeler then got a call from one of the ReChem workers asking to meet him, and he agreed to meet them in a hotel outside the area. "There were three of them . . . the stories they regaled me with; using axes to burst open the PCB barrels, near accidents. . . . They were worried that they hadn't been informed about all the dangers, this was the real problem. I was amazed. Then they said they were going to speak to the *Falkirk Herald*."[41]

On 18 May 1984 the *Falkirk Herald* ran "Living in Fear" by their reporter Alan Crow, which told the workers' story. Around the same time SCOTTIE were organizing the petition and collecting signatures in Stenhousemuir, Larbert, Denny, Bonnybridge and in Camelon. "[in] Larbert," said Wheeler. "I think we must have got about 99 per cent. We probably got one refusal for each sheet. Some of them wanted to sign it in blood. The whole place was in uproar."[42]

It was the prelude to the European elections and the MEP, Alec Falconer, became involved. When the petition was ready they found they had collected 16,000 signatures. We checked those signatures carefully. I know there were some people from Canada, who were visting; their names had to be extracted because the worst thing you could do is give them ammunition. So it was all set out that the vast majority, 16,000 people, had signed that petition."[43] In July John Wheeler received a letter from George Younger's office saying that the college reports and the Lenihan investigation would answer questions relevant to the call for a public inquiry.

The Scottish environment minister, Michael Ancram, said he had not ruled out a public inquiry but said he had no evidence to justify one at present. Within a month Greenpeace joined the call for an inquiry after it had announced the discovery of dioxin in cattle around Bonnybridge. Michael Ancram, they demanded, should hold a public inquiry under section 96 of the Control of Pollution Act, 1974. The following day, on 17 September 1984, as ReChem announced the closure of their Roughmute operation, Alex Falconer called for a local public inquiry under section

96 of the Act.

Ancram repeated his belief that a public inquiry into ReChem's Roughmute operation was not necessary in a letter to Falconer at the end of January. The colleges and Harwell reports had been published and Ancram said the Lenihan report was imminent. Two months later Falconer attempted to rally support for a public inquiry in the European parliament but his resolution failed to get a majority. On 14 February 1985 the Lenihan report was published. Professor Lenihan said there was a need for a more specialist investigation into the babies born with eye defects in the Bonnybridge area.

Calls for a public inquiry in the early months of 1985 continued to be ignored. The *Stirling Observer*, in a strongly worded editorial, backed the community. "The *Observer* is not a sensationalist paper. We simply feel we have a public duty to persue the truth."[45] Euro MP for Mid-Scotland and Fife, Alex Falconer, and Falkirk MP, Dennis Canavan, had consistently backed the community's call for a public inquiry. It prompted them to employ the chemist, Jonathan Wills, to prepare a comprehensive chronology of events, which Canavan presented to Westminister and Falconer to Strasbourg "The government has been absolutely disgraceful in its complacency on the whole matter," Canavan said, and Falconer: "Our evidence shows that central Scotland is not a healthy place in which to live at the moment."[46] Michael Ancram acknowledged that he had received Wills' dossier and had taken note of it but said:

> I do not think that taken individually or together they introduce any substantive evidence which would justify a change in the government's view, that in the light of the report of the Review Group chaired by Professor Lenihan there are no grounds to justify the holding of public inquiry. As you know, and as has been stated many times before, the setting up of a Review Group was widely publicised and the Group invited comments and submissions of evidence, so that local residents (including former members of the ReChem workforce) had every opportunity to make their views known. The Review Group found no unusual morbidity in the area, but made a number of recommendations for further action, all of which are being followed up.[47]

Throughout 1986 questions prepared by a team, which included SCOT-TIE, FoE (Scotland), Jonathan Wills, Alex Falconer and his assistants Richard Leonard and Danny Cepok, were put to the appropriate government ministers in Westminister by Dennis Canavan. He tabled about a hundred questions without proper response. Alec Falconer:

> I've since discovered that it was a breach of what we would call civil rights and civil liberties in Britain, [that] the ministers can get rid of a question by turning around and saying the information could only be supplied at disproportionate cost. So they don't answer the question. [If you ask], for example, how many tonnes of PCBs were brought into Britain? The minister will say, well we could not supply the answer to that question because of disproportionate cost.
>
> We've got to question our democracy and our democratic rights in Britain and how best the public can actually control their own rights. Ministers have a habit of just not answering . . . directly.[48]

In June 1986 John Mackay of the Scottish Office said all the further studies recommended were underway "I do not consider that additional epidemiological investigation is called for at this stage. I have accepted the recommendations of the Lenihan report and am pursuing those which call for further action." In the same month, in response to the question "To ask the Secretary of State for Scotland, when he will publish, in detail and also in a form accessible and intelligible to the non-scientific general public, the results of investigations by his officials and others into the possible presence of dioxins, furans and other toxic compounds in emissions from municipal incinerators," Michael Ancram answered:

> I understand that the tests relating to their incinerators have been carried out on behalf of the City of Edinburgh District Council and Falkirk District Council, to whom the reports were submitted. A report on investigations by the Warren Spring laboratories into the possible presence of dioxins, furans and other toxic compounds in emissions from a number of municipal incinerators in Great Britain is expected to be published during the course of next year.

When pressed for the preliminary results from the Warren Spring laboratory, Canavan was told that trace amounts of dioxins and furans *were* produced in municipal incinerators. He was also told when he asked for a list of areas in Britain with the ten highest and ten lowest background soil levels of PCBs or TCDD that the minister was not aware of any comprehensive survey of soil levels of PCBs or TCDD in Britain.

> The accurate analysis of (dioxin) and (furan) levels in soil is a difficult and costly procedure. My department is currently supporting research work to define the analytical procedure for the extraction, separation

and subsequent quantitive determination of those chemicals in soil. Once this has been achieved, soil samples will be analysed from various sites selected to establish the range of background levels of (dioxins) and (furans) in groundwater. In view of the extremely low solubility of these chemicals in water, significant contamination is unlikely and this is not considered an appropriate medium to monitor.

When Canavan asked if any action was proposed "to ensure that random levels of fat from beef on sale to the public in central Scotland are analysed for traces of chlorinated halogens, dioxins and furans in parts per trillion; and whether he will publish the results of such analyses in detail." MacKay replied, "The Working Party on Organic Environmental Contaminants in Food was set up by the Steering Group on Food Surveillance on 11 October 1985, and its preliminary work has begun. The need for guidelines on such contaminants, in relation to food in general and milk in particular, will be considered in the light of the advice to be given by the Working Party and of other research in this field" MacKay added that the Food Surveillance would look at "various aspects of the toxicological significance of PCBs, (dioxins) and (furans)".

The Scottish Office also revealed, in answer to a question tabled by Canavan on the number of unexplained deaths of livestock between 1982 and 1984, that 511 reports were made to the State Veterinary Service for the central Scottish region. "Unless suspected, tests for poisoning are not conducted," Mackay added.[49]

Why did ReChem pull out of central Scotland? Many observers of the Rechem story believe it was because of the Roughmute workers' complaints and demands during the last months of the operation. ReChem could defend itself against the campaigners but could not defend itself against its own workforce, who knew what went on inside the factory gates. The workforce said as much when they spoke to John Wheeler and to the *Falkirk Herald*. In a secretly-arranged interview in a local hotel which lasted three hours, workers told reporter Alan Crow they would rather be on the "dole" than carry on working under the "present conditions" at ReChem. On the day-to-day running of the plant the workers said that spillages were common and that minor blasts and explosions were regular occurences. They said if they complained about safety standards they were branded as troublemakers; in an area of high unemployment a man sacked by ReChem would not find another job easily. The workers said they did not know what they were burning and knew nothing about PCBs. "We used to split open the metal containers with picks," one worker alleged. "The PCB went everywhere. It used to splash over our overalls and bodies. We thought nothing of it. We treated it like oil. The company never warned us. We only found out what we were dealing with when a shop steward looked up the meaning of PCB in a medical dictionary." Another worker said they wanted "good, effective and above all independent monitoring".[50]

In the same paper the ReChem general manager in Roughmute said the workers had always known what they were handling. Later, in their response to Wills' dossier, ReChem stated that the workforce had been "well aware of the hazardous nature of PCBs, prior to 1984". Written safety instructions, the company said, had been issued to all workers from June 1975. "Revisions dated February 1976 and March 1977 also mention the dangers of absorption and the need to avoid skin contact. All incinerator operators were therefore informed of the dangers of PCBs and were issued with proper equipment. The medical evidence has clearly demonstrated that there were no adverse health effects related to PCB exposure of any kind."[51] Only seven of the 42 ReChem workers were given health tests after they had refused to handle PCBs. ReChem did not complete these tests because they closed the plant down.

When the Royal Commission on Environmental Pollution published their report in December 1985 it stated that it shared the concern of the HWI that, with the closure of ReChem's Roughmute process, "there may soon be insufficient and adequately distributed capacity to burn all the chemicals for which incineration is the BPEO, with the consequence that more will be consigned to less appropriate disposal such as landfill".[52] As ReChem had the only incinerators in Britain capable of taking solid and liquid PCB waste, the closure of Roughmute reduced that capacity by half, yet there had been no shortage of PCB contaminated waste. In 1990 Dr Alastair Hay estimated that the quantity of PCB waste that has already been destroyed is "a very small fraction of the total". Most of the PCB waste, he said, would not come onto the market for destruction until the mid 1990s and would run well into the year 2005. That is why, he said, so many companies are looking at the incineration market; "you've only got to look at ReChem's profitability – it's fantastic".[60]

As has already been noted (see Chapter 3), ReChem's profits increased tenfold during the latter half of the eighties. ReChem's decision to close Roughmute for "financial reasons" must be seen in the light of the fact that they have a virtual monopoly on PCB waste (one of the reasons why Cleanaway expanded in Ellesmere Port was to compete with ReChem in the PCB trade). Turnover in 1985 was £5.8 million, in 1986 it was £6.7 million, 1987 – £9.2m, 1988 – £13.3m, 1989 – £19.4m, and in 1990 – £21m. In 1990 ReChem announced that it had entered into an agreement with the Italian waste management company, Ecodeco SpA, to take an option to invest in a jointly owned company which would run an incinerator in Italy. ReChem will invest 25,000 million lira in the project if Ecodeco can obtain a suitable site by the end of 1993.[54]

It was John Wheeler's belief that the reason why ReChem closed their Roughmute operation "was because of the workforce".[62] Alec Falconer believed it was "the cumulative effect of public pressure and worker lobbying, together with the political pressure, that closed the plant". He added:

> There was no local authority, or any kind of elected representative who would give [ReChem] any support whatsoever, and in my view quite rightly. They kept an awful lot of information from the public and every time you asked them, God almighty, the amount of lawsuits that that company has thrown at people to try and keep them quiet. At the end of the day I don't think it does the company any credit and I think we've learned the lessons from that.[64]

Notes and references

1. *Times*, 18 September 1984.
2. Wheeler, John. Interview with Fiona Sinclair, 1991.
3. See *Scottish Sunday Mail*, 17 June 1984, *Falkirk Herald*, 10 August 1984, *Scotsman*, 25 August 1984, *Glasgow Herald*, 25 August 1984, Greenpeace (London) 16 September 1984 and *New Statesman*, 21 September 1984.
4. See *Scotsman*, 25 August 1984, *Glasgow Herald*, 25 August 1984, Greenpeace (London) 16 September 1984, *Times*, 18 September 1984, *New Statesman*, 21 September 1984, *Guardian*, 20 December 1984, ReChem's response to the Bonnybridge dossier, November 1985.
5. *New Statesman*, 22 November 1985.
6. For the full account see ReChem's response to the Bonnybridge Dossier.
7. *New Scientist*, 22 November 1985.
8. Ibid.
9. Wheeler, J.
10. *Health & Safety at Work*, May 1985.
11. Phibbs, Sian. "The Public and the Polluter – a local perspective of

ReChem's plant at Pontypool", Working Paper no 467, University of Leeds, October 1985.

12. Wheeler, J.
13. Williams, Fiona. "ReChem" *Chemistry in Britain*, June 1985, vol. 21, no. 6.
14. Kerr, Andrew. Friends of the Earth (Scotland), 16 August 1985.
15. Eggleton, A.E.J. "The environmental significance of dioxin, furan and PCB levels measured in the vicinity of ReChem's Roughmute plant" Harwell, AERE G 3394, 14 January 1985.
16. ReChem, "Harwell tests give ReChem clean bill of health", 17 January 1985.
17. *Observer*, 20 January 1985.
18. SCOTTIE. Response to ReChem release of Harwell report.
19. Lloyd, O., Peacock, T.C. and Williams, F.L.R. "The environment and cancer", *Environmental Health* – Scotland, Summer 1985, 1, 17.
20. Hay. A. Interview with Robert Allen 1990.
21. Smith, G.H. and Lloyd, O.L. "Soil pollution from a chemical waste dump". *Chemistry in Britain*, February 1986, vol. 22, pp. 139–141.
22. See "Talking Points", *Chemistry in Britain*, 1984, vol. 20, p. 877; Smith and Lloyd, *Environ. Toxicol. Chem*, 1986, 5; Smith and Lloyd, *Chemistry in Britain*, 1986, vol. 22, 139–141; Eduljee, *Chem Br*, 1986, 22, 308–309; Lloyd, O. et al, *Br Journal of Ind Med*, 1988, 45, 556–560; Lloyd, O. et al, *Chem Br* 1987, 23, 31–32; Eduljee, G. et al. Chemosphere, 1986, vol 15, 9–12, 1577–1584.
23. Eduljee, G "Soil pollution at Bonnybridge: ReChem's reply" 1986, 22, 308–309.
24. Lloyd, O. *et al*, "Bonnybridge revisited" *Chemistry in Britain*, 1987, vol. 23, pp. 31–32.
25. Ibid.
26. Ibid.
27. Wheeler, J.
28. Falconer, Alec. Interview with Fiona Sinclair 1991.
29. Farquhar, S. Interview with Fiona Sinclair 1991.
30. Wheeler, J.
31. Farquhar, S.
32. SCOTTIE, letter to M. MacDonald, Director of Environmental Health, Stirling County Council. 17 June 1976.
33. Clark, R. Interview with Fiona Sinclair 1991.
34. Farquhar, S.
35. Farquhar, S.
36. Wheeler, J.
37. First SCOTTIE report, December 1976.
38. Wheeler, J.
39. Ibid.
40. Wright, I.W.W. (Scottish Office), letter to John Wheeler, SCOTTIE, 6 March 1984.

41. Wheeler, J.
42. Ibid.
43. Ibid.
44. Scottish Home and Health Department, Bonnybridge/Denny morbidity review: report of independent review group under chairman, Professor J. Lenihan, February 1985. Scottish Office Library.
45. *Stirling Observer*, 6 March 1985.
46. Falconer, A. and Canavan, D. Press conference transcript, August 1985
47. Denny/Bonnybridge questions and answers to House of Commons, October 1985, June 1986 and October 1986, from Alec Falconer's office, tabled in parliament by Dennis Canavan.
48. Falconer, A. Interview with Fiona Sinclair, 1991.
49. Denny/Bonnybridge questions.
50. Crow, A "Living in Fear", *Falkirk Herald* 18 May 1984.
51. ReChem response to Bonnybridge dossier.
52. Royal Commission of Environmental Pollution, 11th Report, December 1985.
53. Hay, A. Interview with Robert Allen, 1990.
54. ReChem annual accounts, 1984-1990. See also *Haznews*, no. 28, July 1990.
55. Wheeler, J. Interview.
56. Falconer, A. Interview.

Chapter 6

LEIGH ENVIRONMENTAL, LANSTAR, MBP, CAIRD, WIMPEY WASTE, BLUE CIRCLE

Leigh Environmental

"We have never been run out of any town yet and as long as we are making money out of it we're not going to be run out of Walsall," Leigh Environmental chief executive Malcolm Woods told the *Financial Times* in February 1989. Walsall Community Action against Toxic Waste (WCATW), were determined to force Leigh to close their Stubbers Green operation.

It was during the summer of 1989 that Walsall's GPs began to voice their concerns about the methods of toxic waste disposal. At the annual British Medical Association conference, the Walsall doctors demanded that an inquiry be set up to determine whether human ill-health was a consequence of toxic waste disposal. The Walsall communities had complained incessantly to their GPs about sore throats, burning eyes, skin irritations. The GPs knew how to treat their symptoms but nothing more. The BMA ordered an inquiry.

The communities in the West Midlands had given Leigh a hard time in 1989. It was in 1989 that Dudley Metropolitan Borough Council won their case against the Department of the Environment which had granted Leigh subsidiary, Waste Incineration Services, a licence to burn radioactive hospital waste at their Peartree Lane site in Netherton. The High Court ruled that environment minister Nicholas Ridley had been wrong to ignore the views of the council. Waste Incineration Services had been issued a licence by the council in 1985 which stipulated that no radioactive waste could be burned without council permission even if a government licence was granted.

The community opposition to Leigh's plans for WIS began in October 1988 when the company installed a new incinerator. Leigh had asked the council for a licence to burn 25 tonnes of waste per day and for the time restriction on their operation, which only allowed them to burn waste between 7.30 am and 5.30 pm, to be lifted. The council said no, it would be a danger to public health and detrimental to public amenities. Leigh appealed to the Department of the Environment and then applied to the DoE for a licence to burn 60 tonnes a day, 24 hours a day, seven days a week. They also applied to burn agrichemicals and pharmaceuticals.

The Community Campaign Against Radioactive and Toxic Waste (CCARTW) was formed and began to liaise with the group in Walsall. Increased activity at WIS, they had stressed, would give Leigh a "monopoly on people's suffering". CCARTW said they objected to the siting of the plant in the middle of a densely populated area with seven schools within a mile radius. "It's where it is that we strongly object to," local activist Carol Baker said. "When the wind is in one direction the emissions actually fall on the school playing fields."

CCARTW had made their presence felt with several petitions. In December 1989 a 7,500 name petition was passed onto the Minister for Environment and Countryside to urge government to rule in favour of the local council and reject Leigh's application. Leigh's Malcolm Woods revealed that the incinerator had been operating at one tenth of its capacity and that local opposition had affected the company's profits. In January 1990 Dudley council announced that the Secretary of State had granted Leigh's application to burn 60 tonnes a day. The application to burn agrichemicals and pharmaceuticals was refused. No extra chemicals could be burned and the requested extension was refused.

Leigh are one of the few waste disposal companies in Britain to be successfully prosecuted. When they were fined £500 and ordered to pay £1,495 for breaches of their licence at Peartree Lane in 1989, it did not surprise the community around Leigh's Killamarsh works in Derbyshire. In 1979 Polymeric Treatments Ltd, a subsidiary of Leigh, bought the Killamarsh site and began to use it to store wastes and to treat tar and oil contaminated water. In 1980 the existing boiler was converted to burn contaminated waste solvents. In 1982 Leigh were prosecuted for emissions from the boiler. In 1983 the operation was expanded to recover waste paint solvents. The waste disposal licence was extended in 1985 and permission was granted to remove PCBs from transformers and capacitors but this ceased when the council rescinded the permission. In 1986 Leigh were prosecuted again. They were fined

£1,500 for polluting sewers with their effluent.

Leigh continued to expand their operation, stretching the original planning permission for the site to its limits. In September 1986 the local community formed Killamarsh Residents Action Committee (KRAC) following a fire and explosion which had started in a building which contained old aerosol cans. Fumes from the fire drifted over the nearby housing estate and 14 of the 50 firemen who fought the blaze were given medical checks. A protest march against Leigh's activities ended with a demand for a public inquiry. In March 1987 the DoE refused the inquiry. Leigh were subsequently fined £1,500 for failing to ensure the safety of their employees at the Killamarsh site.

The community, however, were still unhappy about Leigh's operation. Solvent smells had become a regular problem. In April 1988 the regional pollution inspector attributed the smells to Leigh and closed the processes they were doing at the time. The community had continually complained about the smells. They said the emissions made food unpalatable, forced them to remain indoors and made them ill. The Killamarsh community is adamant that Leigh should be shut down. Their fears about Leigh's activities intensified in 1987 when the community learned that the company had purchased the high temperature incineration plant from Berridge Incinerators of Hucknall. The local Parish Council does not agree that Leigh should be forced to close. Leigh employ between 70 and 80 people in a depressed coal mining and steel making district. "We have never taken the view that this firm should be closed down," the Parish Council chairman Bob Harper said in 1988. "We have said we want to be reassured that the company is operating safely. We want to take the fears of the residents into consideration too."

In Doncaster such considerations were not part of the equation when Leigh announced they had plans for the Kirk Sandall site in the South Yorkshire town. Kirk Sandall village was purpose built by glass manufacturers Pilkingtons in the nineteen twenties. During the summer of 1988 a redundant waste treatment site was purchased by Leigh. The site had been established by the four Metropolitan authorities (Sheffield, Rotherham, Barnsley and Doncaster) which made up South Yorkshire County Council. The restructuring of the metropolitan counties left the site in the control of a residual body who were forced to sell it to the highest bidder. Doncaster could not afford to buy it. Leigh was the only bidder, and announced they would operate the site under the same conditions which had applied to South Yorkshire County Council. The local community, alerted that Leigh had bought the site, decided to keep an eye on things. "We did not oppose the planning permission," Liz Jeffress noted because the terms had not been changed. "We did, however, resolve to keep a close watch on things. We convened a meeting with Edenthorpe, our neighbouring Parish Council; it was resolved to closely monitor the situation and to issue a press statement drawing attention to our concern, explaining why we had not opposed the application." Jeffress said they were ready to act if (of when) things changed, and they did a few days after Christmas.

> On Friday evening, January 6 1989, I received a phone call from a fellow Parish Councillor, informing me that he had heard from someone who worked in the planning department at Doncaster Metropolitan Borough Council that Leigh had informed the authority of their intention to put up a 200 feet high chimney on the site to burn toxic waste, and what should we do as the information was supposed to be confidential. I did not feel that confidentiality should apply to an issue like this, so told him to pass the word around and tell people of the January Parish Council meeting for the following Wednesday.
>   I was still surprised to see about 200 people waiting outside the hall on the Wednesday. Not all of them were from the Kirk Sandall area. Our meeting room is very small, with an annex; we managed to cram about half of the people in, but the rest waited outside. Many questions were put at this meeting and quite obviously a lot of fears aired. We decided that the best thing to do was to wait until the application had actually been submitted before doing anything. The clerk was asked to write to the director of planning and a further meeting was called.

Approximately 1,500 people attended the next meeting, at the Assembly Hall in Denton's Green Lane in Kirk Sandall. The parish council borrowed PA equipment to relay the proceedings of the meeting to an overspill in the car park. An action group was set up and a committee elected. Councillor Pat Mullany was elected chairperson. Martin Gregory, a district and parish councillor for the neighbouring ward of Edenthorpe, was elected vice-chairperson. A petition was organized. A local art student designed several logos for the campaign.

At the meeting Mullany told the crowd that Leigh regarded the payment of fines as a relatively minor cost of doing what they intended, making private profit from public nuisance. He said he did not want an environment like the one Leigh Environmental would provide and requested anyone with opinions or concerns to write to him for inclusion in a report to be submitted to the planning sub-committee.

David Ellis, director of planning for the Doncaster council, told the meeting that Leigh had submitted an EIS as part of their planning application. He said the council would accept observations and comments until the end of March. Ellis was asked if the decision making responsibility could be taken out of the hands of the council. He told the meeting it could if Leigh asked the Department of the Environment to call in the application. This, Ellis said, would take the issue away from the council. It would be advantageous to Leigh, but there would still be a public inquiry.

Leigh announced that their new plant would cost £15 million, process 60,000 tonnes of waste per year and create 80 jobs. They insisted there would be no significant environmental impact on agriculture, residents or sensitive industries. Leigh also said they expected that planning permission would be refused but they would appeal to the Department of the Environment and a public inquiry would follow. "Almost immediately, members began asking for a march," Liz Jeffress recalled of the meeting when the Kirk Sandall Action Group was formed. "This was planned to coincide with the planning meeting and we went ahead on that basis." However, the planning meeting was postponed. The campaigners decided to hold the march anyway.

> We took our petition with us and marched the four miles to the Mansion House, where the Mayor (councillor Ray Stockhill) accepted the petition from Pat Mullany. The march was led by Norman West, our MEP. The three Doncaster MPs, Mick Welsh (Doncaster North – the constituency in which the site lies), Martin Redmond (Don Valley) and Harold Walker (Doncaster Central) marched with us, in the body of the procession not at its head, as they respected our wishes that we wanted party politics kept out. We had full co-operation from the Labour Party; many of the leading campaigners are Labour Party members. The Tories kept a very low profile. The Liberal Democrats were active. The Greens wanted to be a part of the technical sub-committee but only as observers; this was not agreed to by other members of the committee so we didn't see too much of them. The Socialist Workers tried to muscle in on us at the beginning but we felt we had to see them off.
>
> Several firms on the (Kirk Sandall) industrial estate gave their workers time off to go on the march. Frigoscandia, the company most likely to be affected, closed down, and all their employees, factory floor and management, marched with us.

The planning meeting was rescheduled for June. The campaigners obtained yellow balloons and ribbons and invited protestors to bring their cars, suitably decorated in the campaign colour. Over 200 vehicles took part in the motorcade. The council turned down Leigh's application on the grounds that their proposed plant would be a threat to the local community, industry and water supply. In the Doncaster racecourse exhibition hall, where the majority of the campaigners had gathered, there was, as Liz Jeffress described it, "a sigh of relief" when the decision was announced. The planners had heard that 300,000 people lived within a 10 mile radius of the site, with Doncaster town centre two miles away. Over 60 industrial units surrounded the site. Ten schools are located within a mile, with the nearest home merely 300 metres away. The councillors were adamant that Leigh should not get approval. "If Leigh were to come here it would be the kiss of death to the revival of Doncaster. Thanks to this government Britain has become the dustbin of Europe but Doncaster isn't going to become the dustbin of Britain," councillor George Brumwell told the planning committee. Councillor Bob Granger had said: "The water table is only five metres below ground level and if that becomes contaminated Doncaster would have polluted drinking water. The experts tell us that would be almost impossible to rectify." After the meeting Leigh director Edward Wilkinson told the media his company had "provided a plan for the safest and best facility in the UK and we can't do better than that". The protestors braced themselves for a long winter campaign before the public inquiry.

They engaged a barrister to prepare their case and represent them at the inquiry and he decided they should concentrate on three key areas: hydrology, toxicology and the environment. "The technical sub-committee sought out the expert witnesses and we had several meetings prior to the beginning of the inquiry to ensure our case was presented in the best possible manner," Liz Jeffress recalled.

Two pre-inquiry meetings with the inspector took place, one in December 1989. We emphasized that the public were not entitled to take part, and that we had to impress upon the inspector that we were "responsible" people. About 500 turned up to this meeting and the only time the audience made its presence felt was when the inspector announced he would visit the continent in order to observe existing incinerators. The Leigh representatives suggested that the inspector might like them to make arrangements for him. There was a low murmur which grew into a very loud roar of "NO", and a cheer when the inspector said the other parties could go at the same time as him but he would not allow them to accompany him, and that his travel arrangements would be handled by the Foreign Office.

The week before the public inquiry we again had the caravan in the town centre and our object here was to push up the signatures on the petition to 100,000. We needed 3,000 signatures on the last day and didn't think we would make it, but we did achieve that target by lunch time and in fact managed over 5,000 at the close of day. The start of the public inquiry was again a beautiful day, a real carnival. Frigoscandia provided a decorated low loader, the bus drivers from South Yorkshire Transport decorated a bus, the vintage car club joined us and another local company sent a low loader for the jugglers who had volunteered to take part. I was in one of the front cars and as I got to the Dome where the inquiry was to be opened the police told me that the last car had just left Kirk Sandall, five miles away.

Leigh had only engaged their QC two weeks before the commencement of the inquiry. During the whole of the inquiry he didn't seem all that bothered. On the last day of the inquiry the junior counsel for Leigh told me that he came from a mining village in the Welsh valleys and knew exactly how we felt when our community was threatened and wished us luck.

At 4.55 on the evening of Monday 11 November 1991 Liz Jeffress heard that the government had decided to turn down Leigh's appeal on the grounds that the operation of the incinerator would affect the local water supply.

The inquiry had heard about Leigh's environmental record and the fines resulting from their Killamarsh operation. Leigh, however, did not let that worry them and they continued with their plans for expansion. But it did worry the people of Cadishead, Davyhulme, Eccles, Partington and Salford in Greater Manchester when Leigh submitted, in April 1990, a planning application to Trafford Park Development Corporation to build toxic and clinical waste incinerators in Nash Road, Trafford Park. According to leigh, the proposed plant would be "the most modern and technically advanced facility in Europe which will be available to solve many of the region's environmental problems".

Trafford Park in Greater Manchester was the first, and some would argue the greatest, purpose-built industrial estate. It is still the largest industrial complex in Britain and Ireland, possibly in Europe. In 1970 the Trafford Park Industrial Council was formed, acccording to Davyhulme MP Winston Churchill, to ensure "the continual development and prosperity of the estate as a major industrial and commercial centre, and of improving its environment". In 1970 the death rate from bronchitis in Manchester was the highest in Britain. Air pollution from Trafford Park was the major contributor. By 1990 the pollution had been drastically reduced. Respiratory illness was no longer seen as a major problem. The community wanted it to stay that way.

Leigh's proposal immediately met strong opposition from the local community and from the food companies in Trafford Park. The proposed site was across the Manchester Ship canal from a large housing estate in Eccles. Over 38,000 signatures were collected on a petition to oppose Leigh's plans. Approximately two thousand people compiled personal objections.

Erica Woods, one of the campaigners and a co-ordinator of Community Campaign Against Toxic Waste (CCATW), recalled that it was just by chance that they saw the small advertisement which announced the proposed development in the public announcement columns of the *Manchester Evening News*. "We read those columns, because if you missed that public announcement the permission would sail through as nobody would have said a word against it."

The 5th of April was the first announcement that (Leigh) wanted to build a clinical waste incinerator – 15,000 tonnes a year – and a toxic waste incinerator for 25,000 tonnes a year – there were to be PCBs included in that. The difficulty was in making local people realise what a potentially dangerous situation this was. We created a lot of fuss and bother. We contacted the food companies and we contacted all the companies within the vicinity of the proposed incinerators. Initially it was Green Party members, then a multi-party campaign was launched at a public meeting in Salford in June [1990]. A lot of local people turned up for that meeting, about three hundred I think. They all started to write letters to the press and to the [Trafford Park] Development Corporation expressing their concern. As the concern spread Trafford Park DC arranged public consultation meetings so that Leigh Environmental would have an opportunity to explain their plans. There were two meetings; one was held on the Salford side of the canal and the other on the Trafford Park side. At the first meeting we thought we had an unanimous "no" – there was one lady at the front who voted yes, she apparently liked the idea, everyone else voted no. Apart from four young people at the second meeting it was all no as well. I suppose about a hundred attended.

The Stretford Labour MP, Tony Lloyd, sought and won a debate in the House of Commons where he also rejected Leigh's plans. He was supported by the Conservative MP, Winston Churchill. Lloyd said he was pleased to have the support of Churchill "because I wish to establish from the outset that this is not a partisan party-political issue but one that unites the entire community in our area". Lloyd said he wanted to raise the matter "because my own son and three daughters live in the area in which it will operate. As both a parent and a constituency member, I am concerned about the impact of that incinerator on the health of my children and those who live in the area". He added:

> I do not want to be accused of the NIMBY syndrome but I feel that we should recognize that, although the disposal of PCBs and other chemicals must be dealt with, we have no obligation to take on the world's problems. We should think seriously about what policies we adopt towards the importation of waste. Even in their unincinerated form PCBs are dangerous chemicals and in the Stretford and Davyhulme constituencies in Manchester the lorries that carry PCBs are themselves dangerous, which in itself worries my constituents.

Lloyd criticized the role of the HMIP. They showed incompetence, he said, and expressed concern that the HMIP should have no plans to block the proposal, subject to planning approval by the corporation.

> At public meetings held recently in Manchester, the representatives of Leigh Environmental were asked how they would respond if there were unacceptably high levels of emissions of dioxins or other chemicals from the plant. They said that nothing could be done for days. Worse still, the inspectorate does not test for such chemicals but simply accepts the results provided by the company. It is unacceptable for it to be left to the company to provide evidence about whether it is causing pollution.

He also criticized Leigh and quoted from a company log book, which he said was compiled by company operatives. "A company which, according to its own operatives, puts the safety of its equipment and personnel secondary to production is unfit to operate a plant in a residential area." He alleged that Leigh had a disastrous record and said: "An inefficient plant is an incredibly dangerous source of pollution."

Churchill said he wished to re-emphasize the "deep concern and disquiet" that was felt by his Davyhulme constituents, but he added that the public concern about the proposals extended to Eccles and Salford, immediately across the canal, and "to the entire Greater Manchester area". He said that although the government had a responsibility to secure the safe disposal of chemical wastes "the heartland of a conurbation of three million people is not the proper place for that to be done". He said Leigh had admitted, at a public meeting in the Old Trafford cricket ground, "that PCBs and dioxins would remain undestroyed". Leigh, he said, had claimed the levels would be low;

"but that factor cannot be ignored," Churchill added.

> The incinerator is to be sited in Trafford Park, upwind of the conurbation of Greater Manchester, with prevailing westerly and north-westerly winds, and will have a devastating effect throughout the area downwind of the plant, which includes the centre of Manchester and surrounding dormitory suburbs. Local residents, are not the only people to be concerned. Industry and commerce in Trafford Park feel that their interests will be blighted, none more so than the food manufacturers. One must ask, when HMIP, seemingly with great and indecent haste, rubber-stamped this proposal, subject to planning permission by the Development Corporation, did it consider the implications for the food processors in Trafford Park? Kelloggs is in Trafford Park, producing cornflakes for the entire nation, and other major food processors are there. Is that a suitable place for a toxic waste incinerator, alongside major food-processing companies?

Thirteen major food companies with factories in the industrial estate agreed with the two MPs and sent a delegation to London to meet Environment and Countryside Minister, David Trippier. A spokesman for the delegation said: "People must have safe and wholesome foods. Therefore, there should be laws to ensure toxic waste disposal plants are never sited near food operations because unique considerations apply. Toxic waste processing is a comparatively new industry and its safety record, both in the UK and worldwide, does not inspire confidence. Even with the most modern plants and sophisticated monitoring equipment, accidents still occur." CCATW said, in one of their newsletters: "UK laws and European Community guidelines prevent food operations being sited near sources of possible contamination. Similar restrictions should prevent toxic waste disposal plants being located near food operations."

On the day that Trafford Park Development Corporation unanimously rejected Leigh's application the companies launched a campaign to press government "to introduce legislation to ensure that plants to incinerate toxic wastes are never located in areas where large numbers of people live and work or where there are food operations". Colin Doeg of Brooke Bond and also spokesman for the 13 companies said: "The affair has exposed a lack of adequate laws about this sort of situation. We are not against high-temperature incinerators in waste disposal, but we don't know much about their impact, so they should be sited in safe areas where they can be controlled and monitored." It has been estimated that Trafford Park industry spent £250,000 in their campaign against Leigh's proposals.

Following Trafford Park Development Corporation's rejection of the proposals, the company said they would consider an appeal after it had studied the reasoning behind the rejection. Leigh had until the end of February to appeal and CCATW warned: "The fact is that Leigh have the right to appeal to the Secretary of State within six months of the decision. If Leigh appeal there will be a public inquiry. If we lose the case at the public inquiry we will get the incinerators." The campaigners' fears were certainly justified. In reply to Lloyd's and Churchill's demands that Leigh's planning application be withdrawn (before Trafford Park DC made their decision) Robert Atkins, Under-Secretary of State of the Environment, said: "I am not persuaded that a sufficient case has been made for taking the exceptional step of calling in the planning application. I understand people's fears about the proposal, but I do not think that fears alone, unsubstantiated by evidence, are sufficient to justify calling in the application."

By the end of 1990 Erica Woods did not think Leigh would back down, but she said the community would be ready for a public inquiry, which she said was designed to intimidate the layperson. "It will be taken very much out of the hands of ordinary people. All we can do is gather as much information as possible to be able to feed a recognised qualified bod with information so that he (sic) can interpret our case; we're going to have to pay out a lot of money for professional representation. It will have to be raised from the community. No one else is going to give it to us."

Leigh decided not to appeal the decision. The deadline – 24 February – arrived and went. One campaigner, Jane McNulty, said: "It is not beyond the realms of possibility that [Leigh] could put in a fresh proposal at some time in the future – we are ever vigilant!"

The campaigners in Greater Manchester did have an advantage over the Doncaster campaigners. They already lived in an area where pollution was a major problem. Jane McNulty explained:

In April 1970 hundreds of gallons of a highly volatile substance was accidentally discharged into the Ship Canal from the Shell Chemicals tanker dock on the south bank. An early morning passenger on the local foot-ferry unsuspectingly threw his cigarette into the water and caused an explosion and fire that claimed his life and that of seven others on the ferry, badly injuring another two. The flames rose 60 feet in the air and spread the width of the canal for three-quarters of a mile. I was 14 at the time and remember well the panic that ensued. If the wind had been coming from the south that day, Lanstar (or Lancashire Tar Distillers as it was then) would have gone with it, for sure, and probably so would have Cadishead.

Highly dangerous substances are transported through our town daily by road, sea and rail. We live in a triangle of chemical reprocessing plant and landfill tips, the greatest density in the country I believe. Lanstar is just one, albeit the greatest polluter of the area. No one knows what substances are being dealt with at Lanstar and its probably just as well. In November 1989 we were all confined indoors following a gas cloud leak. In 1988 the same thing happened – a total of 13 people hospitalised.

Road tankers, containers, drums continue to arrive and leave, some under the cover of darkness; the stench is ever present. Manchester Ship canal is one great, dead sewer. But the wages are good and employment is precious here. The councillors and MPs dare not say too much.

Lanstar

Lancashire Tar Distillers was set up in 1929 as an oil recovery operation. During the eighties LTD became part of the Lanstar Group and a fully licensed multi-purpose plant was developed in Cadishead for the recovery, treatment and disposal of liquid industrial chemicals and effluent by Lanstar Waste Treatment and Wimpey Waste Management in a joint venture. Lanstar Wimpey Waste had, according to one of the company's management, "an extensive waste treatment and recovery facility" which could treat "a wider range of wastes than any plant in the UK". LWW recovered valuable chemicals, treated drums and prepared waste for incineration and landfill.

Campaigns against the polluting industries along the ship canal have, according to Jane McNulty, always existed. "Funnily enough there was never an outcry in 1970 when the great fire happened. You'd have thought if anything was to be done it would have been done then, but there was nothing. Maybe it was the shock and the grief. Then there was a small spill [in 1988]. That was enough to get people angry and asking questions and that was when Partington and Cadishead Environmentalists (PACE) was formed."

The people of the area had always accepted the smells and the respiratory illnesses. "Washing used to disintegrate on the lines. One night, in the mid-seventies, a cloud from Lanstar blistered the paint on all the cars. They actually paid us compensation for that. I don't know if it was ever documented but a neighbour of ours had his car re-sprayed courtesy of Lanstar. At the same time, when my daughter was a baby I couldn't leave her in the pram in the back garden because of the stink that came over from Lanstar," Jane McNulty said.

People have taken the smells and the incidents for granted – they all complain; but they tolerate it because that's the way life is. People will say the air is a lot cleaner than when the steelworks were in operation. It's true because I remember when I was a kid you'd get up in the morning and there would be red dust all over your window-sills. It was red oxide dust coming from the chimneys. With the steel works it was obvious. You don't know what pollution from Lanstar would look like but just because you can't see it doesn't mean it's not there. People have looked at the licences and have looked more closely at the operations and have complained to the environmental health authorities. Why people are more concerned now I don't know; there isn't that much difference between the stuff which came out in the past and they stuff that is coming out now – it's all toxic.

In the mid-eighties, when the Lanstar Group began to expand its activities in the area, a campaign was launched after the company bought land to the west of its site which they wanted to use as a dump. That group was the genesis of the group now known as Irlam and Cadishead Hazards Initiative (ICHI), which grew out of a campaign against Lanstar's importation of toxic waste during 1990. When the story broke that Lanstar had a contract to take Italian toxic waste from the *Zanoobia* (see below, Chapter 7), over 250 people turned up for a public meeting which had been hastily arranged by word of mouth with a few posters displayed in local shops. "At the end of the public meeting people were asked to volunteer for a committee and 50 people put their names down so 50 people were on the committee," Jane McNulty said. "They met two weeks later; we decided to have a march and to leaflet all 3,000 homes in Cadishead."

About 450 marched from Irlam swimming pool to the Lanstar factory in September 1990. "It was planned to disrupt the traffic. We did it at evening rush hour and held up traffic for miles in both directions. It was quite a media event and we got the evening TV news and radio," said Jane McNulty. A petition which had been organized by a local woman was handed into Lanstar. The campaigners said their aim was not to jepordize jobs in Lanstar but to stop the importation of toxic waste and to ensure the highest safety standards in the treatment of indigenous waste. With 440 employed in Lanstar there was a genuine fear among the workers and in their trade unions that jobs would be lost. One shop stewart told ICHI that he was very concerned about the way safety procedures in Lanstar were ignored. "His members are less concerned, he despairs of their apathy," Jane McNulty was told. The unions' thoughts about toxic waste were one thing, she was told, "but the actual members don't see it as a problem".

Lanstar's management had said they would attend the public meetings which were being held in the area. At one meeting they admitted that they had imported toxic waste from Italy, but said they had the technology to deal with it, and that the import was only paint residues and solvents.

One local reporter, who had been investigating Lanstar's activities from their back fence, discovered that the workers actually did believe the company's explanations.

"What are you looking at?" a worker demanded.

The reporter said he was looking at the barrels of toxic waste. "Do you know what you're working with here?"

"Oh no, there's nothing poisonous here," the worker replied.

"You're standing next to a barrel that's marked toxic."

"Yes that's toxic not poisonous."

When the march took place the Greater Manchester WDA admitted they had given LWW a licence to dispose of 271 tonnes of waste from the *Zanoobia*.

In November 1990 Charterhouse Venture Funds, who had invested in the ReChem buy-out from British Electric Traction in December 1985, announced they had put £4 million into Lanstar. Josiah Lane, the major shareholder in Lanstar, was replaced as chairman of the company by Dr John Walker of Charterhouse. Lane said he expected Lanstar to remain an independent company and predicted a successful future. "Lanstar has invested substantial capital in new plant, which places it in a leading position to respond to the demand for safe and efficient waste management and oil recovery, following new environmental legislation."

In May 1991 Lanstar made four men redundant which the campaigners attributed to the new machinery. In contrast to 1990, 1991 was quiet. The principal activists in the community groups discovered that the interest had gone. "The intensity of peoples' anger is brilliant but short-lived, their interest-span lasts only as long as the threat is visible: out of sight out of mind," Jane McNulty lamented. "It will take another near disaster or scandal to make them sit up and start protesting again."

ICHI's membership had remained high, but from regular attendances of around 50 some meetings during 1991 were inquorate. Jane McNulty had seen it all before:

> The problem with groups like PACE and ICHI is that they don't have specific aims. The aims were too broad and general. We had public meetings that were very poorly attended – half a dozen. We invited speakers. We had Hilda Palmer from the Manchester Hazards Centre. We had the National Rivers Authority and the Waste Disposal Authority

sent George Clapton. They assured us that they had their eye on the situation, that things were in control, that the rule book was being applied strictly and there was nothing to worry about. Actually the membership of PACE accepted it. There were 20–25 people involved in the group, eight on the committee and about 25 people came to the meetings. After the initial outcry about the solvent spill and the original public meeting where people brought out their complaints in an open forum – there were anecdotes about the apple tree that died, all that sort of thing – they seemed quite content to go away and not do anything else about it. When the speakers came, I think people were quite relieved to be assured there was no problem. I dropped out [of PACE] when the chair of the meeting said: "Obviously there is no problem at Lanstar – its just very unsightly; could we not write to the management and ask them to plant a line of conifers at the back so we don't have to look at it." I couldn't believe that attitude.

Motherwell Bridge Products

The plans for expansion by the major waste disposal companies have not been confined to England and Wales. In Scotland, "an interesting little market", as one company spokesman described it, proposals for incinerators, treatment plants and landfill sites marginally outnumbered the mergers and takeovers, in the early nineties. Caird, Wimpey Waste and Shanks and McEwan were joined by Motherwell Bridge, Lanstar and Blue Circle. Anti-toxic campaigners like Fiona Sinclair and Foster Evans are convinced that Scotland has been set up to become Europe's largest toxic dump. The trends for waste disposal which evolved between 1988 and 1990 in the central and Highland regions and the deletion of planning legislation by the Scottish Office, they claimed, would leave Scotland extremely vulnerable to the importation of toxic waste. In January 1990 Fiona Sinclair expressed her concern about official monitoring of toxic waste sites. Scotland has roughly the same number of waste disposal authorities as England and Wales, where the population is times larger.

> In effect, this means that expertise with regard to waste monitoring is sacrificed to breadth rather than depth of knowledge. This problem is compounded by the fact that Scottish local authorities have 100 vacancies for environmental health officers. Waste monitoring is carried out with the District Council, which is also responsible for waste disposal, whereas the functions of waste monitoring and waste disposal are split between the Borough and County Councils in England and Wales.

The activity, which saw Motherwell Bridge move into Scotstoun, Caird into Renfrew and Wimpey Waste into Balmullo, has disturbed the campaigners who are convinced that the waste disposal industry can see a future in Scotland. The campaigners are unsure whether they should believe the industry's claim that they are in Scotland simply to dispose of the country's own waste. Motherwell Bridge Products (MBP) had no doubt that is why they wanted to move into Scotstoun.

In 1988 Crown immunity from prosecution was removed from Health Board clinical waste incinerators with a capacity above one tonne per hour. New legislation, which came in during 1991, removed immunity from incinerators with a capacity of less than one tonne per hour. The Health Boards reviewed their disposal arrangements, as many hospital incinerators would not meet the new air pollution standards. They were faced with the prospect of investing in new technology or finding alternatives for clinical waste disposal.

In this light Motherwell Bridge Products saw a potential market for a central clinical waste incinerator. During the summer of 1989 they decided to build a clinical waste incinerator on a site they owned and which was zoned for industrial development. MBP proposed to burn 21,000 tonnes a year – a capacity of 57.6 tonnes per day; the output of the seven central belt health boards in 1990 was 46 tonnes a day. MBP said they would not accept industrial, commercial or municipal waste. The Glasgow city planners said the site could be considered if an environmental impact assessment was done. MBP commissioned an EIS from Environmental Technology Consultants Limited (ETC) of Newcastle upon Tyne.

Harland Cottages residents were the first to hear about MBP's plans. They were notified because they would be neighbours of the proposed plant. In a letter dated 8 September 1989 the residents were told that the application was for an "incinerator which would deal with hospital waste, the waste being delivered in sealed skips daily and the incinerator being housed in a standard industrial building. . . " The letter went on to state that the application was "not for a hazardous waste incinerator and the latest technology means it will surpass the new standards laid down by government and the EEC, without any hazard to those working in or living next the plant".

The Harland Cottages residents, being quite isolated, tried to fight the incinerator proposal on their own, so it was quite a while before the community realized the significance of the proposals. It wasn't until March 1990, when MBP were invited to speak to a public meeting, organized by the local Labour party, that the significance of MBP's proposal hit the community. The meeting was packed out – two halls were used – and about a thousand people attended. ETC, Motherwell Bridge and George Galloway (MP for Hillhead, which covers the Scotstoun area) and Environmental Health officers from Glasgow District Council all spoke.

The proposal had provoked strong local opposition. The major concern was the effects on the health of the community, the effect on amenities in the area, and on property values. Disparate groups of people began to work on their own in different campaigns. A further public meeting in March formed the basis for the present non-political group, and Campaign against Clydeside Incinerator (CACI) was formed.

Cathy Russell, who had been elected press officer of CACI, immediately sent out a press release. The following week she went into the Planning Department. "It wasn't really 'til I read the application myself that it hit me what we were dealing with. When I read that this incinerator could burn the waste from 190 hospitals, I just realized that this was something way out of scale to anything that had ever been built [in Glasgow]."

Nina Baker, the secretary of CACI, wrote to all the community councils, to get city-wide support. The first week after the Easter holidays, CACI called a press conference, "and that's when things really started to move", according to Russell, "especially when George Galloway volunteered to lie in front of the trucks!" The conference went well, but they were disappointed with the *Evening Times*, who had only given the campaign a small piece. "So, I wrote to the *Evening Times*, explaining what the proposal was about, enclosing a campaign briefing note, and the next day I got a call from the editor of the *Times*, saying 'I'm going to back your campaign'." When she told her fellow campaigners the good news, she didn't feel that they fully appreciated the significance of it. "There's a world of difference between getting your stuff reported, and actually getting editorial backing."

CACI decided to commission their own report on MBP's proposal from scientists and experts living near the proposed site. Two health and safety officers had been elected to the CACI committee. The council also commissioned a report. Planners, environmental officers and a representative from Aberdeen University were invited to vist the site in Dudley.

The CACI technical team got out their report in three weeks, greatly aided by Motherwell Bridge's own EIS, which they pulled apart. When questioned as to whether they had challenged the company on their specific proposals, or on incineration *per se*, Mr X (a government civil servant) said: "we weren't just saying that incineration by itself was a

bad idea, but particularly incineration on the site they were proposing."
Ms Y (a government civil servant) elaborated, "they could put the same
incinerator in another area, which isn't heavily polluted and has a
different geographical layout, different topography, and wouldn't
affect anybody's health." MBP, they said, had chosen the site simply
because they owned it.

Three members of the group had knowledge of incineration with
regard to the law, but Peter Malcolm, another member and senior
lecturer at Kilmarnock College, had a more detailed knowledge on
the chemical effects. CACI submitted their technical report to the
planning department. The campaigners claimed there were mistakes
and significant omissions in MBP's report. CACI went to MBP's
headquarters and delivered letters from children – a primary school
is located less than a thousand yards from the site; a map was also
delivered by the objectors showing the proximity of housing to the
proposed incinerator.

Mass rallies took place and objectors were invited to address a
sub-planning committee before a decision was made. Sandy Simpson
of MBP admitted that the opposition was stronger than he expected,
but "once it is up and running they will realize it's just a factory
next door."

CACI had a strong team, the committee of 24 people had a good
grasp on the relation between science, politics and economics, and they
were able to criticize MBP's report for its inaccuracies. Radio Scotland's
'Speaking Out' programme gave the issue an hour and people were
encouraged by CACI to phone in and "give Sandy Simpson [of MBP]
a really hard time". The programme subsequently won a British Radio
Award as the best environmental programme, and *Evening Times*
reporter Ally McLaws was commended in the Natura awards for his
coverage of the anti-incinerator campaigns in Scotstoun and Renfrew.

By 15 May 1990, when the campaign in Renfrew began to surface,
the Planning Department stated that it would recommend the turning
down of the MBP application when it went to committee. "That was
actually quite difficult for us, because an awful lot of people read
that article and thought it was all over," said Cathy Russell. CACI
organized a rally, which gave a focus for their political lobbying. Clyde
Port Authority got in touch, after being sent a copy of the CACI report.
The rally did not attract any media coverage. MBP's plans went to the
planning committee on the 28 June. The city planners ruled that the site,
in a densely populated area, was unsuitable for incineration. MBP said
they would appeal to the Scottish Secretary. CACI said the battle would
only be won when the Scottish Office said "no". The campaign had, at

this stage, all-party support.

The Public Health Department at Glasgow University produced a report which described the proposals as "an experiment with the health of the local population". Dr Graham Watt's report considered the potential health effects of the proposed incinerator. "It is not possible to provide a quantified assessment of the risks due to emission of heavy metals, or of low level radioactive waste, or of dioxin. With low exposure levels the risk of such emissions would be likely to accrue over the long term. It would be very difficult both to detect adverse health effects and to attribute them to specific environmental hazards."

On the emissions from the proposed incinerator, Watt stated: "It is important to note that guidelines have been established for single chemicals. Chemicals, in mixture, can have additive, synergistic or antagonistic effects." Breach of SO_4 limits could occur as constant low levels, breaching the yearly average, or as peak emissions, breaching the one hour or 24 hour limit. The WHO report on "Air Quality Guidelines for Europe" stated: "It's not clear whether long term effects can be related to annual mean values or to repeated exposure to peak values". The proposal was that the incinerator operate continuously and the company's only legal obligation would be to ensure emission levels at or below legal limits. It seemed possible that the gases combined with background levels could breach the 24 hour limit.

Hospital incinerators are difficult to operate satisfactorily because of the different wastes burned, and the different combination of wastes burned at any time. The planning application had shown no evidence that the incinerator would achieve the legal emission limits. Watt deemed this scientifically unacceptable and stated that no treatment in the health service would be accepted on the basis of such incomplete data. "If no information is available about the performance of the incinerator which is proposed . . . it would be appropriate to . . . recognize its initial operation as an experiment."

The planning application had stated: "On the basis of the plume dispersion modelling, taking into account the strict standards to be imposed by current regulations on the quality of the flue gas, it can be demonstrated that the proposed development will pose no adverse impact to the health of anyone living or working in the vicinity."

Watt could not accept this. It was based on assumptions that the emission levels would be below legal limits and that legal limits are safe. The total level of air pollution was not taken into account. There are separate EC limits for gas emissions from incinerators and for general air quality. The guidelines do not provide a guarantee of safety below the legal limits.

Special-risk groups must be considered, as they would be more susceptible to the effects of pollution than the general public. Those most at risk would be pregnant women, children, the elderly and people with chronic respiratory illnesses. At any time there are likely to be at least three hundred pregnant women in the area. Within the area there are 1,380 children under school age and over two thousand over school age. Within 1 km of the proposed site there is one secondary school, three primary schools and one nursery school. There are only two thousand people in the area over 75 years of age.

Watt concluded his report with the recommendation that the incinerator, with its uncertain risks, be sited elsewhere. It was, he said, an experiment with the health of the population. Glasgow Health Board had no comment to make on the report's findings and conclusions.

A spokesman for the health board had said in December 1990 that they had two options: "The first is to build our own incinerator on an existing hospital site, to dispose of all our clinical waste. The plan involves the heat from the incinerator being used to supply energy to the hospital. The other option is to contract out to Motherwell Bridge, but the costs would have to be competitive and we would have to find a suitable site first." CACI said they favoured the health board's option for their own incinerator. "At least that way we will not be at the mercy of a merchant operation which could branch out into the burning of a wide range of hazardous waste other than hospital waste." Until an option is chosen Greater Glasgow Health Board still have to spend approximately £1 million upgrading their existing incinerators.

Caird Environmental

The campaign against Caird's proposals for an incinerator in Renfrew began after the community learned that Cleveland Fuels Limited, a wholly owned subsidiary of the Caird Environmental Group, had bought the 10 acre Buchanan Oil site at Meadowside, Renfrew, complete with incinerator. The licence to operate the existing works had expired on 1 April 1988 and Renfrew District Council were opposed to Caird's plans. Local people thought that the incinerator was going to be for clinical waste, because they knew that the Greater Glasgow Health Board was looking for sites for these. In March 1989 more than two thousand residents signed a petition to protest against Caird's proposals to expand the incinerator to burn various wastes, which the company later revealed would include chlorinated solvents, fuel and waste oils, and oily sludges.

Foster Evans got involved with the issue, as he was chair of the local Labour party, and they ran the initial campaign – they made a presentation of the petition to the district council on 1 June. The council unanimously accepted the petition and agreed to monitor the site. Caird chairman, Peter Linacre, had said they would not burn toxic waste, but only oil-based wastes and that these are the same types of chemicals that have been treated and disposed of at that site for 60 years." Foster Evans, who later became a co-ordinator of the campaigning group, Clydeside Against Pollution (CAP), said: "This should not be located in a settlement the size of Renfrew and we will fight all developments which threaten the environment."

In May 1990 Ally McLaws highlighted Caird's plans for the site in a two-page feature which stated that Scotland was set to become Europe's fastest developing toxic waste dump with giant incineration plants and landfill sites. Caird, he wrote, would apply to burn 10,000 tonnes of "residues from all forms of industry and maybe sewage sludge" as soon as they completed the refurbishment of the incinerator. Caird's development director, Kevin Bond, told McLaws that the incinerator would be brought up to EC safety standards before operations began. Renfrew councillor Archie Driver said: "There will be strong local opposition to the re-opening of the incineration plant. I am always suspicious of big waste companies and their motives." McLaws reported that the old incinerator had been the subject of scrutiny by the pollution inspectorate. The community, he wrote, had breathed a sigh of relief when it closed in 1989.

In a telephone conversation with Foster Evans, Bond claimed Caird had been misrepresented. He did not feel the article was a fair reflection of what they were trying to do, and said that they had proposals, which he could not let him see then, but there would be a public presentation of them later. Bond came across to Foster Evans as very credible.

On Friday 13 July 1990 a notice appeared in the *Glasgow Herald* advertising the application for registration of the incinerator. But the notice had appeared during a summer holiday period, a point made by Renfrew District Council in their representatiion to the HMIPI in Edinburgh on 2 August 1990. Because the council feared that people potentially affected by the incineration works might not have seen the application, it was suggested to the HMIPI that "the representation period should be extended or the advertising procedure should be repeated so as to permit proper consultation and representation". (Foster Evans saw the petition in the local newsagent.) The notice pertained to Caird's application for registration by HMIPI. The documents did not arrive in Renfrew District Council's offices until four days after the notice had appeared.

The campaigners got in touch with various councillors and community groups on both the Renfrew and Glasgow sides of the river, making sure that Renfrew District Council put in a response in time. The campaigners brought the matter to the attention of Glasgow District Council, who, because of the 21 days allotted for objections, had not had an opportunity to discuss the proposals in committee, although they had previously discussed Caird with the Environmental Health Department. (The campaigners and the District Council have since won a concession from the Secretary of State for Scotland, that all submissions, irrespective of when they arrived, will be considered.) The Labour party then called a meeting, where they decided that "this was too big an issue to be party political", and a community group should be set up. This was something which Foster Evans was insistent on, because of the involvement of people who were not involved in politics.

The first public meeting was called in September in order to set up a campaigning group; about three hundred and fifty people attended, in Renfrew Town Hall. A subsequent meeting in October chose the name of the campaign, Clydeside Against Pollution was chosen because of the involvement of people from Yoker, across the river, and from Paisley. The death of a local MP brought a by-election, and CAP decided to stage a rally, during which every candidate stood up and supported the campaign.

In a five-page document Renfrew council, the WDA for the area, outlined its position on the application. Among a series of points, the council stated that the new works would be capable of emitting pollutants including dioxins from the incineration of PCBs and chlorinated substances, which would have a significantly greater environmental effect "than those previously experienced arising from the process at J.O. Buchanan" and that

> the plant is located within the Clyde valley and emissions from the stack will not only affect the inhabitants of Renfrew District but also those of Glasgow and Clydebank Districts. The area itself is densely populated and industrialised and consequently suffers a high pollution level from a variety of sources. The ability of an area to tolerate any new pollution source is dependent on the existing level of atmospheric pollution. This factor has not been taken into account by the applicants, who have assumed in preparation of their dispersion model that zero pollution exists and therefore that the emissions from the incinerator, as predicted, will not exceed current guideline limit values. The council request that HMIPI take prevailing background levels of pollution into

account when determining the acceptability of otherwise of emissions from the stack.

The council document concluded with the comment that there was "sufficient serious concern" to oppose the application.

Earlier the regional chemist expressed his concern about the application, that it contained "serious omissions", notably that "pollution from this works has been assessed for the commonest weather conditions but not for the weather which would produce the worst levels of pollution". He said his own assessments "show there would probably would be complaints of offensive atmospheres from the public, and there is the possibility that health limits would be breached". He added: "As a result of these considerations I consider this application should be rejected."

The community were totally opposed to Caird's incinerator. In October Scottish television estimated that nearly two thousand people marched to the site. In January 1991 another petition, with 8,000 names, was collected and sent, along with 9,000 postcards, to Ian Lang at the Scottish Office. A third public meeting was attended by over a hundred and fifty people.

By the end of January the campaigners expected a decision from the Scottish Office. "The Secretary of State for Scotland will be advised by HMIPI in Scotland whether to grant Caird the right to register their proposed incinerator in Renfrew. A decision was expected in November 1990, then January 1991. We are advised now that they still can't make a decision and don't know when they will," CAP announced in their newsletter which had been sent to community councils, groups and approximately three hundred and fifty businesses. Renfrew council, said CAP, would still make Caird apply for planning permission and a hazardous waste disposal licence if the incinerator was approved by the Scottish Office. "If it gets the go-ahead then we shall continue the fight against any planning permission being granted. It is therefore important that we keep the pressure on the Scottish Office and maintain the support of our local councillors."

Caird, meanwhile, had not remained quiet. In a letter to *The Scotsman* at the end of January, the chairman of Caird (Scotland), Norman Semple, said it was a sad paradox that green protestors so often cause the very environmental damage they seek to prevent. Not far from Renfrew, he wrote, "there are old-fashioned incinerators doing an essential job in burning clinical waste and other wastes, but causing untold pollution. . . . So we have the crazy picture that councils, driven by objectors, are doing all they can to stop new facilities being built and operated to current highly-regulated (and very expensive) standards while at their very backs old ones are the problem". He concluded: "Furthermore, any company serious about staying in business cannot afford to get a bad name by any sort of non-compliance. It just would not pay."

In April 1991 CAP surveyed seven local GPs and sent an open letter to Caird and Severn Trent Water, the newly-privatized water company who had made a £78 million bid for the Caird Group in September 1990. CAP said they had no confidence in the "financially troubled" Caird Group. "Our campaign is to stop Caird's proposals. The scheme is flawed, wrongly located and potentially dangerous."

The following month Foster Evans told local reporters that six of the seven GPs surveyed had agreed that the "construction and operation of the incinerator would be detrimental to the health and well being of the people of Renfrew".

Questions from Eileen Adams, the local MP, to the Scottish Office have elicited the response that only steam will be produced. The Scottish Office has also told local MPs that the objections being made are really planning matters, and as such, should be dealt with at the planning stage. Noise and location are put forward as planning matters – but the campaigners believe that location is not, in this case, solely a planning matter, because of the recognized high levels of pollution in the area.

Caird has also been asked questions by the campaigners, which they have not answered. Some of these pertain to the investigation of Caird's subsidiaries by the Fraud Squad, Inland Revenue and DTI.

Foster Evans also believed that the MPs have taken a higher profile in the Scotstoun campaign. The Strathclyde East MEP, Ken Collins, wrote a letter to Evans, saying that, with the EC Directives coming out, there will be no international transfer of toxic wastes. Foster Evans quoted from Collins' letter: "So far as the import and export of toxic waste is concerned, the European Community is rapidly moving to a position where the movement of waste from one country to another will be prohibited. It is very unlikely, therefore, that when this legislation is enforced, toxic waste will be moved into any site in Scotland, or indeed anywhere else, by sea."

Wimpey Waste Management

Scotland has a problem with the disposal of its toxic waste. In the eighties much of the country's toxic waste was landfilled, the rest sent to England and Wales. Of Scotland's 835 waste disposal sites, 700 are landfill. In 1990 the Scottish Office admitted that it did not know how many landfill sites received toxic waste. Official figures showed that approximately 38 thousand tonnes of toxic waste is sent each year to Paterson's, the main landfill site in Scotland, which is under licence from Glasgow District Council. Both the environmentalists and the authorities in Scotland agreed that this was not satisfactory. It was this market which Wimpey Waste spied.

In April 1987 Wimpey Waste Management were granted permission to restore and infill Brackmount Quarry near Balmullo with inert waste such as builders' rubble. The following year Wimpey Waste applied for permission to deposit commercial, industrial and domestic waste. It was claimed they could not develop their business in Fife until permission was granted. The local community lodged objections with the planning department of the North East Fife District Council. Permission was refused. Wimpey Waste lodged an appeal with the Scottish secretary, Malcolm Rifkind.

Villagers from Leuchars, Guardsbridge and Dairsie presented Rifkind a petition calling for a public inquiry. They feared the build-up of methane gas in the dump and the objectors believed leachate would contaminate Motray Water and the Eden nature reserve. The hillside into which the quarry has been gouged is covered in wells, there are 27 ground springs on the floor of the quarry, and, as the quarry drains in different directions, it empties out into a large number of burns that feed one of the most fertile areas of farmland in Scotland. There is a faultline running through the quarry, and the layering of the rock is such that, according to one of the objectors, "migration of gases would carry the gas maybe 200 metres, right up to the nearest habitation." This information was provided in surveys commissioned by the objectors. Heavy traffic, the objectors claimed, would endanger the north sea gas pipeline 250 yards from the site. RAF Leuchars, a mile from the site, wrote to the Scottish Office to express their concern that the increase in birds at the tip would endanger its military operations. Guardsbridge Paper Mills said it was worried about its workers and its water supply.

A public inquiry was held in Cupar in August 1989. Dr John McManus, a geologist at St Andrew's University, represented the community. The local authority, the RAF and the nature conservatory were also represented. Guardsbridge Mills withdrew their objection before the inquiry ended and conditions were agreed with Wimpey Waste.

William Halkett, a local Green Party activist, attended the inquiry. "I'd been to one or two other planning enquiries, and I've seen them seeming to go remorselessly against the evidence. In planning inquiries, there is always a predisposition in favour of the developer, that's the way these things work, but in this case it was so blatant from day one, that I could hardly believe it", he explained. Opposition to the dump had mainly been focused through Balmullo Community Council.

"If you were trying to conduct a study of the place where you would not dump wastes, this is it," he added. "The other thing about it was that Wimpey waste turned up with absolutely nothing [apart from] a QC. They'd no site survey, no seismic survey, no hydrological survey. They'd no development plan for the site and no management plan. In fact, what they actually said, with regard to the surveys, was that it was an 'intolerable burden'. It was unfair, UNFAIR of anyone to expect Wimpey Waste to provide these expensive surveys, when they might just as easily be turned down."

The Inquiry Reporter agreed with the company about this, and their contention that they were not obliged to provide this material, and could leave it until the site licence stage (when the public no longer has any say in the matter). "What they actually did was to play off the two stages against one another, and whatever they were asked, they said, 'Oh, we will do that at the site licence stage'."

According to William Halkett, Wimpey Waste's "scheme just evolved on the spot, and it started from, as I say, absolutely nothing, and became the most elaborate landfill scheme that I have ever heard of". The company first put forward a plan to line the quarry with clay, and when it was pointed out that the springs in the quarry would make this impracticable, they put forward plans for an "impermeable membrane" of plastic, which they thought would last for about twenty-five years, but which has only been in use in Europe for 10–15 years. "A flare stack could be provided" for burning off gaseous emissions from the landfill, there would be "total exclusion netting", to try and avert problems with birds. Leachate would be collected by tankers and taken from the site, but nobody thought to ask where to. All these ad hoc plans convinced Halkett that "the only way that that quarry could ever be viable as a landfill site is if they bring in special wastes, they can't make money any other way." Questions of the cost of these plans were never asked, nor was such information volunteered.

Wimpey Waste's appeal was upheld by the Scottish Office. Inquiry Reporter, John Henderson, said the concerns raised by the various parties opposed to the development "relating to property values, visual amenity, smells, dust, vermin and tree planting" do not "carry sufficient weight to justify refusal of planning permission". He said these problems could be prevented by good site management. "I am satisfied that the risk emanating from escaping gases are acceptable in planning terms," Henderson said.

The Reporter's letter of 22nd November 1989 on the inquiry was contradictory: "The commercial and industrial elements [in the waste] would be non-hazardous. Radioactive, toxic or tankered [liquid] wastes would not be deposited. A reasonable definition of toxic was 'hazardous to health'. It was not possible at this stage to be precise about what exactly what would be deposited nor to say for definite that 'special wastes' would be excluded (paragraph 6)." Special wastes are, under British regulations, the most toxic of all toxic wastes. The Reporter ruled: "By refusing the application on grounds of potential pollution, the Planning Authority had abused its powers. A full Environmental Assessment had not been requested by the District Council and could not be insisted on in terms of the recent regulations." This was said despite the upholding of the need for an EIS by the Secretary of State for Scotland, for plans by the same company for a landfill site in Renfrew District, which resulted in Wimpey Waste withdrawing its application for planning permission.

Why had North East Fife District Council not pressed for an EIS, and why had their Planning Department recommended conditional approval of Wimpey's plans, which were rejected by the councillors? The council also gave, as one of their objections to the dump: "The adverse impact of your client's development, would be chronic and more drastic because the detection and rectification of a leak would be difficult." In view of the nature of the site, as previously described, it is not hard to see why.

The area's Westminster representative took a good deal of criticism from local residents. Menzies Campbell MP "had previously acted in a legal capacity for Wimpey Waste Management", but did "not understand that planning permision has been given for "toxic" waste at Brackmount Quarry". When pressed on the question of his continued employment by Wimpey Waste Management, he had "nothing to add" to his earlier statement, but it later transpired that he was still acting for the Wimpey Group.

"When Guardsbridge Paper Mills finally withdrew their objection, after they'd cobbled together a deal with Wimpey Waste, the Reporter then turned to the rest of the objectors and said 'Well, is that it?' He continued 'Now that the main objector has withdrawn its objection, does anyone wish to continue?' and, of course, the enquiry room went bananas, people were leaping up and down, shouting," recalled Halkett. There was no verbatim account made of the proceedings, which Halkett believed was a major drawback, not least because the account written by the Reporter "does not give you the actual spirit in which the thing was conducted".

In May 1991 the environmental health department of North East Fife District Council stated they believed Wimpey Waste had not sought to use the quarry for the disposal of hazardous or toxic waste. The director of environmental health said Wimpey Waste had made an application for planning permission, but had not submitted an application for a waste disposal licence. He added that planning permission does not cover those types of materials. None of the landfill sites in the district were licensed for hazardous or toxic waste disposal. "Indeed all such waste is exported from this district for specialist disposal."

Blue Circle

The private sector is never slow to exploit new markets. The new legislation which affected the disposal of clinical waste had attracted the waste disposal industry. So it was when the bovine spongiform encephalopathy (BSE) scare brought changes in the legislation governing the disposal of carcasses. Many traditional knackers, who had collected dead animals from farms, went out of business and the agricultural industry needed an alternative method of disposing of about 3,500 tonnes of dead animals annually from farms on the Scottish and English border.

John Salvesen, a farmer at Spylaw, Kelso, near the border saw a potential market for carcass disposal and planned to build a £500,000 carcass incinerator. However, feasibility studies showed that an incinerator could not run profitably on one product. Salvesen concluded that he would increase the range of products to be handled at the proposed plant and he was approached by Blue Circle Waste Management. Following consultation with Blue Circle the original plans were revised. Salvesen and Blue Circle Waste Management lodged an application with the Borders Regional Council to build a £2 million waste incinerator on land owned by Salvesen Farming Partnership near Hunter's Hall, Kelso.

The proposed incinerator would handle animal carcasses, medical wastes, industrial and trade wastes, special wastes and document disposal. The Environmental Impact report by Salvesen's Kelso Waste Ltd and Blue Circle Waste Management stated: "It is initially intended to bring in some 40 tonnes per day of medical and animal waste." The report said that the total capacity of the plant for phase one will be 45 tonnes per day. It referred to the company's "intention to extend the plant at a later stage (say after one year's operation) and then increase the operation by building a second plant, and probably a third and fourth in the future".

Salvesen said the plant will be "cleaner than clean". He said it would run initially on waste oil but would produce energy from the waste it burned, converting steam to electricity. The report stated: "The ash will not constitute a biohazard as the furnace temperature of 1100 degrees Centigrade will destroy any toxicity" and "the waste plant will not emit plume smoke". Remaining treated waste would go to a landfill site. According to Salvesen the new plant would create about 25 jobs. "We are very central, about 90 miles from Glasgow, Teeside and Cumbria. We could cover it all. I'm very optimistic about planning permission."

There is opposition, particularly among businesses in the area, to the proposal. There is concern that vehicles carrying the toxic waste from the south of Scotland and north of England could be involved in accidents in areas with inadequate emergency services. Seven objections were lodged with Borders Regional Council planning department.

James Murray, chairman of John Hogarth Ltd, a milling firm which supplies oatmeal and barley products to national food producers, said an effort would be made to prevent the plant from being built. The proposed incinerator is to be sited 100 yards from the mill, which employs 40 people. "This incinerator will burn all kinds of stuff and animals that have died from salmonella, foot and mouth disease and BSE will have to be brought along a road which passes within a few yards of our premises. I believe the possible spread of disease from fallen livestock and medical waste to the human food chain would be quite high."

Neve Electronics, who manufacture consoles for television and radio stations, are situated 200 yards from Hunter's Hall and employ 200 workers. Ron Stevenson, works director, said: "We have persuaded our parent company, Siemens, to invest £1.3 million in new equipment at Kelso. It is vital we have a clean environment because electronic components are susceptible to fumes and emissions. We have lodged a formal objection in the interests of our business and our staff." Details of the project were presented to a public meeting in Kelso in May 1991, organized by planning officials. Ian Borthwick, assistant director of planning, said the department would be seeking authorization to appoint a consultant who would advise on the environmental impact assessment provided by the developers.

The concern about toxic waste disposal is widespread throughout England, Ireland, Scotland and Wales. Communities fear new proposals for toxic waste dump sites and they fear they consequences from sites, as we have seen, which have been in operation for years.

Note

Sources for this chapter were personal interviews, press cuttings, correspondence between the communities and bureaucracy/industry, minutes of meetings, planning documentation, including environmental impact assessments, Hansard, company records and personal observations.

PART TWO

THE TRADE AND THE TOXINS

Chapter 7

THE TOXIC WASTE TRADE IN BRITAIN

In 1989, in its report on toxic waste, the Environment Committee, under the chairmanship of Sir Hugh Rossi, concluded "that the import of wastes for treatment or incineration in Britain was acceptable provided it is properly regulated at all stages".[1] Toxic waste imports increased significantly throughout the eighties, the majority of waste being transported to Britain by sea. The International Maritime Organization estimated in 1989 that more than 50 per cent of cargoes transported by sea are hazardous. Nearly thirty ports in Britain handle hazardous wastes.[2]

Importation of hazardous wastes began in the mid-seventies and escalated 15-fold between 1981 and 1987.[3] Between 1981 and 1982 3,800 tonnes were imported. Hazardous waste imports comprised 52,000 tonnes between 1986 and 1987. The largest exporter was the Netherlands (55 per cent) with Ireland and Belgium next (12.5 per cent each). A proportion of the Dutch waste would have originated in other European countries and would have come through Rotterdam. [4] Hazardous waste imports reached 80,000 tonnes between 1987 and 1988.[5] In 1990 waste imports declined by one third. The main sources of imported waste in 1990 were Switzerland (10,327 tonnes), Belgium (5,846 tonnes), Ireland (2,948 tonnes), Italy (2,651 tonnes), Netherlands (2,648), Portugal (1,068 tonnes), Austria (637 tonnes), Sweden (554 tonnes), Spain (483 tonnes) and Germany (445 tonnes).[6]

Toxic waste imports into Greater Manchester Waste Disposal Authority area doubled between 1985 and 1986, from 63,220 to 123,518 tonnes with 24,489 tonnes coming from outside Britain. Between 1986 and 1989, however, the imports to Greater Manchester remained almost constant, with 135,040 tonnes imported in 1989.[8] Officials of Greater Manchester Council believe the main reason for the growth in toxic waste imports is the restrictive attitudes of European governments and local authorities towards hazardous waste disposal within their own countries. The chief officer of the Greater Manchester Authority, Alec Davidson, said: "Our biggest worry is that the industry will grow so large that without proper controls and regulations being imposed on it there will be so many options for misdeeds, the cowboy operators, call them what you like, that there will be virtually no authoritative supervision."[9]

Nearly 19,000 tonnes of waste were imported through Humberside in 1987 and almost 15,000 tonnes were destined for Greater Manchester. Over 3,000 tonnes went to the West Midlands.[10] The vast majority of wastes reaching Humberside in 1989 were transported by road to Wimpey Waste in Manchester – 492 out of 553 loads. Of the remainder, 36 went to ReChem in Pontypool and nine went to Leigh in the West Midlands.[11] Cleveland imported 16,000 tonnes via Teeside in 1987. The waste was mostly Swiss and went to the Manchester area. In the first six months of 1988 the West Midlands were notified of 26,000 tonnes of imported wastes. This was a considerable increase compared with 240 tonnes in 1982.[12] The main destinations for toxic waste imports in 1989 were Lanstar and Leigh in Greater Manchester (16,346 tonnes), ReChem, Pontypool (8,936 tonnes) and Leigh, West Midlands (10,871 tonnes).[13]

Humberside has become the prime recipient of European hazardous waste. Consignments come in through the Humber estuary for unloading at Immingham or Hull docks. Within a two year period the importation of hazardous waste through Hull and Immingham tripled. In 1985, 3,500 tonnes of hazardous waste were shipped to Humberside and this rose to 12,109 tonnes in 1986.[14] Humberside received almost 19,000 tonnes of mostly chemical and paint industry waste from European producers in 1987. The Humber has good shipping links with major European ports. The county has direct links between the docks of both banks of the river and there is a good motorway connection; these factors have contributed to the increase in the flow of toxic waste through the port.[15] The amount of hazardous waste imported through Humberside dropped in 1989 for the first time in five years, 553 shipments compared with more than 1,000 in 1988. The 1989 shipments totalled 13,753 tonnes. They included acid waste, industrial grease and sludge, acidic tar, solvents, used batteries and thousands of used aerosol cans. Most of the waste originated in Holland and Belgium, more than 400 of the 553 shipments. Brian Taylor, Humberside's Waste Control Officer said that the major reason for the reduction was that many other countries were developing their own facilities,[16] and that many contracts, were not completed with brokers shopping around – Cleanaway lost 38 shipments of PCBs to France, who undercut British prices.[17]

Hull's seamen claimed in 1987 that they were being kept in the dark over the toxic cargoes they handle. "We never seem to be able to find out what the stuff is and what amounts are coming in," said Hull's NUS secretary Ken Turner. "By the time we have gone away and consulted the code books, the stuff has gone on its way." He said they were calling for a ban on the importation of waste. Councillor Terry Geraghty of Humberside County Council said: "The people of Humberside deserve to know what is arriving at their ports and being carried on their roads."[18] In 1988 it was proposed that Humberside Hazardous Substances Liaison Panel be set up to act as a forum for co-operation on the transport, storage and handling of dangerous goods within Humberside. The panel was set up with the Emergency Services, Department of Transport, the local HSE, the national liaison officer and the County Council.[19]

Felixstowe handled more than 13,000 tonnes of toxic waste in 1989, one third of the total toxic waste imported into Britain that year. The majority of this was sent to Leigh in Walsall which took 6,100 tonnes, and ReChem in Pontypool which took 945 tonnes. This waste mainly originated in Holland, Luxembourg and Belgium.[20] The largest tonnage of waste passing through Felixstowe comprises PCBs. Three quarters of PCB waste imported into Britain in 1989 went through Felixstowe. Figures show that 4,700 tonnes of PCBs were shipped through Felixstowe in the 12 months ending 31 October 1989. Other toxic wastes imported through Felixtowe were chlorinated solvents (419 tonnes) and pesticide wastes (110 tonnes).

Bill Rampling, county secretary of the Fire Brigades Union, said firemen were worried about the increase in the volume of imported toxic waste and the way it was being transported. "With waste we are often not talking about standard chemicals, but concoctions which may pose unknown hazards." A spokesman for Felixstowe Dock and Railway said it was the responsibility of disposal companies to inform the local authorities of the movements of loads through their areas. "We do not ban any cargoes provided the relevant regulations are complied with."[21]

Until 1987 the small Tees and Hartlepool port complex never recorded imports of hazardous waste. In 1987, through a Swiss connection, the waste trade through the estuary rose to 1,600 tonnes. The county's waste disposal officer Ray Maughan said: "We are powerless to do anything about this potentially very dangerous traffic through our ports and our county." The Tees has good road links with disposal sites in Lancashire and the West Midlands. Most of the waste passing through the Tees in 1987 was sent by one agent – Transamex in Lucerne, Switzerland. The waste was transported by road through Yorkshire, to a site operated by Leigh at Trafford Park. Maughan said the loads consisted mainly of waste and solvents from Europe's paint industry.

Regardless of the port of entry, waste can be transported to any part of the country for disposal. Most of the waste arriving in Humberside goes for treatment or disposal facilities in Greater Manchester, the West Midlands, Gwent and Essex. Pressure of space in port areas means that consignments rarely spend the three days at the dockside required by law to give authorities time to check the loads. Within hours of landing on the quay the waste can be on its way to a disposal site. Most of the waste is driven from Immingham on the M180, M18 and M62 motorways, and the driver of a large heavy goods vehicle can have a consignment across the Pennines in three hours. There are no regulations to hinder the transport of waste and this practice has made Greater Manchester WDA the largest waste disposal authority in Britain. Members of GMWDA have called for stricter controls on importation. A spokesman for GMWDA said: "Because hazardous wastes are mostly shipped in sealed containers from the country of production, it is not possible to establish the accuracy of the consignment notification until the load is opened at a waste disposal site. This means that European producers can export what they like, because by the time the consignment is checked, it is too late to do anything about it."[22]

Transport unions have become increasingly concerned over this growing international trade. The problems associated with the transport of hazardous imported waste were reviewed at a TGWU meeting in March 1990. Drivers and dockers are never informed about what is being carried and sometimes drivers are untrained – the majority of hauliers don't train their drivers in the transport of HazChem. An investigation by the National Road Transport Committee confirmed that the situation was getting worse. Checks in the Dartford Tunnel found that 82 per cent of HazChem transport contravened the law. Toxic waste has been loaded onto a passenger ship, the *Herald of Free Enterprise*. In another case, a Belgian waste broker arranged a deal in which the "disposal site" turned out to be a country bungalow. Because of the increase in chemical loads the role of shipping and forwarding agencies has changed. Shipping agents sometimes act as brokers for hauliers, and the agents often give misinformation to haulier companies and drivers when they come to pick up a load. Some ports have no special facilities for the handling of hazardous waste. Fire Brigades are not informed that chemicals are coming through. Cleveland has the only routing system for hazardous waste. In 1992 cross border checks on freight from the EEC will be abolished at sea ports, airports and all British borders. This has created much concern.[23]

The increase in the importation of hazardous waste to Britain can be attributed to many factors. One reason cited is that British disposal fees have been much lower than those in Europe and the US. Iris Webb, East Anglian spokesperson for FoE, claimed that disposal prices in Britain were about five times cheaper than in France or Germany. "That is why we have had a 20-fold increase in toxic waste imports during the past decade."[24] British regulations are less stringent than those in other countries such as the US, Canada, Germany, Denmark and Sweden. Sweden has put a moratorium on incineration while the technology is being reviewed. British licensed incinerators would be illegal in counties such as Denmark.[25] In Germany paint residues are no longer incinerated following the introduction of more stringent regulations.[26]

Certain types of wastes, classified as hazardous in other countries, would be unacceptable for disposal there. Such waste may not be defined as special waste in British law and is imported for disposal in Britain. In 1987 these imports are thought to have totalled more than 160,000 tonnes.[27] This discrepancy in regulations can be seen in the proposed import of PCB contaminated soil from Germany for landfill in Swansea. The soil was refused disposal in (East) Germany because the PCB levels were unacceptable under German regulations but the same levels would be acceptable by British standards.

Many American cities are considering constructing incinerators but the cost of at least $100 million per plant will raise the disposal price. A number of new incinerators are planned for Germany and the estimated cost of each plant is £200 million. Lack of landfill sites in some countries means waste is exported for landfill in Britain; the Netherlands exported 130,000 tonnes to Britain in 1986, composed mainly of contaminated soil, ash, and industrial waste. The Netherlands has severe problems with landfill because of its geology and high water table.[28] Hazardous waste is imported from developing countries because they do not have the technology to deal with it. Environmentalists argue that it is the multinationals who create these wastes and they should be forced to deal with them at source.[29]

The international toxic waste trade is, as far as the British government is concerned, acceptable, provided that the controls on imported and home produced waste are the same. The Control of Pollution regulations apply to everyone who handles waste in Britain. An importer of waste is the same as a waste producer; effectively the point of entry becomes the source of production.[30] In June 1988, Professor Sir Jack Lewis, chairman of the Royal Commission on Environmental Pollution, wrote to Environment Minister Lord Caithness, expressing the Commission's concern about rising imports of waste. "The Commission suspects waste producers are seeking the easiest options rather than the best," he wrote. The government made it clear, both in answers to Parliamentary questions and in the reply to Sir Jack Lewis, via Lord Caithness, that it believes "trade in waste disposal is a proper and legitimate business provided it is conducted in a manner which is environmentally safe and does not create problems for the importing country. It is sensible and environmentally desirable that particularly difficult wastes should travel to facilities which are capable of dealing with then properly".[31]

The British government has taken the view that a ban on hazardous waste imports may be "resented by developing countries who have a legitimate trading interest".[32] The Environment Committee reinforced this stance. "The future of co-disposal is also important to developing countries. In their situation, where current practice is often open dumping of hazardous wastes in admixture with everything else, controlled co-disposal can be seen as the only practicable option for upgrading standards, as least in the short term. If the UK can be seen to make co-disposal work, then we will have done a considerable service to the international community."[33]

The importation of waste has been defended for economic reasons by ministers. Sir Hugh Rossi, chairman of the House of Commons Select Committee on the Environment, argued: "It is an important contributor to the nation's balance of payments."[34] Chief Scientist of the Department of the Environment, Dr Fisk, also recommended that Britain should play a strong role in the international waste trade for economic reasons. "I think it would be generally true that one would expect an economy like that of the United Kingdom, which has a good tradition in process chemistry and good high quality facilities, to play a major part in [the] international trade in special wastes. Indeed it is a high value added sector and it is part of the service that we can provide where the standards are appropriate."[35] The government opposes imports of bulk wastes of low toxicity, on the grounds that the countries who produce them should be responsible for disposal. However, ministers say imports of smaller quantities of highly toxic waste, such as PCBs, which are difficult to dispose of safely, are to be encouraged. They said that British companies with the technology to deal with them provided a global environmental service, while earning foreign currency.[36]

There is considerable benefit to disposal companies from their involvement in the import trade and such imports contribute to the viability of many special facilities.[37] The National Association of Waste Disposal Contractors expressed strong support for the continued import of wastes for treatment subject to proper and effective controls; they emphasized that the continued viability of some plants such as high temperature incinerators, which are necessary for the efficient disposal of domestic toxic waste, could depend on the continued availability of imported hazardous wastes for treatment. However they did concede that the importation of wastes was an emotive issue.[38] The NAWDC does believe that the EC Directive on Transfrontier Shipments of Hazardous Waste needs strengthening. The NAWDC say any material which could be regarded as waste, whether hazardous or not, should be covered by the directive, and that all cross border waste shipments should have to be notified to the authorities in both countries and accompanied by a full description.[39]

Environmentalists and local communities have objected to the burning of hazardous imported waste. David Wheelar, operations director of ReChem argued: "There are many worse things than PCBs being transported every day". On the NIMBY attitude, he said: "There has been some opposition from local people. We are inviting them to the plant to see exactly what goes on. After all we have nothing to hide. We are on the side of the angels, disposing of these dangerous chemicals so that they don't pollute the environment."[40]

During the summer of 1989 a consignment of waste from the US fertilizer and chemical FMC Corporation shipped to Wath Recycling was examined by the HSE. Wath Recycling believed the waste consisted of no more than copper residues. The HSE analysis revealed that the waste contained high levels of dibenzofurans, toluene and xylene. Wath Recycling were issued with a prohibition notice preventing them from dealing with the waste. An operation to seal the waste into 2,000 steel lined drums began. Three workers, who had unloaded it, were monitored for possible ill effects.

Seven containers of the same shipment were found in a rail depot near Stourton. British Rail workers decided to strike until it was removed. The National Union of Railwaymen said BR had failed to implement safety checks when the nature of the waste was revealed. BR launched an inquiry.

The HSE and the company demanded that the waste be taken back by FMC. The Department of the Environment said the foreign holder of the waste should take it back, but said the government had no power to enforce this. The DoE had contacted the US EPA and scientists were sent from the US to investigate the consignment.

The local community demanded that the waste be removed from the site. A meeting attended by 500 people heard calls for a public inquiry. In December 1989 junior Environment Minister, David Hunt, agreed to meet the Wath Environmental Action lobby, who were pressing for the closure of the plant until the waste was removed. The DoE had said the waste would be removed by 18 December. The community said the government had let them down. A candlelight vigil was held at the factory.

FMC claimed they had sold the waste to Amlon Metals before it was shipped. The documentation did not mention either company. When the waste arrived in Britain the documentation was filled out by Euromet, who were owned by Amlon until 1985. The principal shareholder of Amlon, Israel Weinstop, is also a director of Wath Recycling.

The US EPA investigation revealed that no material classified hazardous under American law had been found in the waste. The campaigners claimed these results had been obtained to protect the US company who had sent the waste. The anger increased when the campaigners learned that the government was doing nothing to force FMC to take back the waste. Some campaigners believed they had a case under US law to claim damages for the mental anguish caused since the waste had arrived in Wath upon Dearne. The mayor of Rotherham wrote to George Bush, but he simply passed the letter to the EPA who reiterated that the waste wasn't deemed hazardous under US law. In February 1990 Labour MP Peter Hardy attempted to bring a drum of the FMC waste into the House of Commons. In May 1990 the High Court granted FMC's application that the case should not be heard in Britain after Wath Recycling, Euromet and Amlon Metals had taken legal action to force FMC to take back the waste and pay damages.

The waste remains in storage in Britain.

Poison cargoes

In the late eighties a series of controversies heightened public awareness of the toxic waste trade. Newspaper headlines "Poison Ship", "Toxic Cargo" and "Cargo of Death" generated opposition to imports.

Dockers and unions became concerned about the dangers involved in the handling and transport of such cargoes. The Zeebrugge disaster illustrated the loopholes in regulatory control of waste importation.

In March 1987, at Zeebrugge in Belgium, P&O Townsend Thoresen directed a lorry aboard the *Herald of Free Enterprise*, which subsequently capsized with the loss of 188 passengers. The lorry was carrying 12 tons of hazardous waste which comprised six drums containing cyanide wastes; the rest of the load was an assortment of highly toxic chemicals which should not have been in the same shipment. The driver, employed by EC Transport, Dorset, objected because on previous occasions he had sailed on freight only vessels. He said his employer had sent a fax requesting the lorry be taken on a freight only ship – the fax stated the lorry contained a consignment of dangerous toxic waste. A spokesman for EC Transport said the lorry should never have been allowed aboard the ship. "We showed those shipping company guys in Zebruggee nine pages of detailed notes on what we were carrying. We'd made 15 previous crossings carrying that stuff and every time we'd requested and been put on a freight only vessel. Our truck was also carrying those special orange warning plates on both the front and the back telling the world we had dangerous goods on."

The port authorities at Zeebrugge denied they were responsible for allowing the lorry on board. "We just check the papers of the drivers who then go to the offices of the shipping company they are booked with. It is the shipping company concerned who tell the drivers which vessel they must sail on," said a spokesman. P&O said they allocated the ship on basis of the manifest presented by the haulier. EC Transport denied insinuations that they had presented an incorrect manifest.

The chemicals were retrieved from the ship after the accident by Alba International of Aberdeen. They were repackaged under the supervision of experts from Harwell and stored at the Belgian army base at Fort Melchen. Their origin was traced to two waste disposal companies: Reingers, Germany and Chinesea, Switzerland. These firms had contacted a Swiss company, Transamex, which had hired EC Transport to deliver them to Leigh, Walsall. MEP, James Proven said: "This business is of the utmost urgency. . . . In a broader sense, we must establish control over the shipment of these chemicals."[41]

In 1979 and 1980 a Canadian company, Cie SOTERC Inc., received a licence to store PCB wastes in a warehouse at St Basile le Grand, Quebec. The company were authorized to store a maximum of 90,900 litres of liquid PCB wastes. The PCB contaminated wastes remained in storage until August 1988 when the warehouse caught fire. At the time of the fire, the warehouse contained approximately 101,800 litres of askarels, a quantity of oils and contaminated residues, stored in barrels, capacitators and transformers. Following the fire 3,800 local residents were evacuated until tests showed the area was safe.[42]

In September 1988, David Powell, a member of STEAM, in south Wales, listened to an HTV telephone interview with Martin Clermont, vice president of the Canadian company International Environmental Materials. He learned that IEM intended to send the dioxin and PCB contaminated remains of the fire at St Basile le Grand, to ReChem in Pontypool for disposal. Powell wrote to Donald McDonald, the Canadian High Comissioner, to object to the export of this waste. He said the people of South Wales did not want their area used as a dumping ground for the waste of other nations. Powell pointed out that the arrangement had been made without the consent of the Welsh public.[43]

David Powell received a letter from the Canadian Ministry of the Environment which said that the export of PCBs from Canada to Britain had been going on for several years but that "the PCB contaminated remains of the fire at the storage facility in St Basile le Grand, however, will not be exported from Canada".[44] Press reports in July 1989 indicated that the PCB contaminated fire remains would be sent to ReChem. Powell wrote again to Donald McDonald, requesting again that the waste not be not sent to Wales.[45] In a reply to Powell's letter, the Canadian High Commission said the decision to send the PCBs was taken because of a processing bottleneck at a Canadian waste disposal facility. The Commission stressed that the exportation had full approval from the British Department of the Environment.[46] A statement from the Canadian government explained why the PCBs were not destroyed in Canada: "The reason is that we don't presently have the facility to do so and one of the reasons for this has been the willingness of plants, in Great Britain, France and other parts of the world, to accept these materials. In other words, the UK has been providing this service and we have been using it."[47]

In August 1989, the *Khudozhnik Saryan* sailed into Tilbury docks carrying a load of PCBs from Montreal. Greenpeace activists stuck a skull and crossbones on the hull. The shipment was turned away by port managers because of "environmental concern." The waste from the St Basile le Grand fire totalled 3,000 tonnes and was to be carried in 15 shipments at weekly intervals. The first cargo was to arrive in Liverpool in mid August. Liverpool port authorities and dock workers said they would refuse to handle the waste and were banning PCBs from Mersey Docks. Initially 38 ports followed suit.[48] Jack Dempsey, docks district officer with the TGWU said: "If dockers have handled PCBs in the past, it has either been without knowing it, or without knowledge of how serious they are."[49]

When the first cargo from St Basile le Grand arrived on board the *Nadezhda Obukhova*, community opposition groups and Greenpeace joined dockers in a demonstration at Liverpool docks. Mersey Docks and Harbour Company refused to unload the PCBs and the ship returned to Montreal. The second consignment arrived a week later aboard the *Khudoznik Pakamov* and was not unloaded. The ship was arrested by the Admiralty Marshall Vincent Ricks, on behalf of Dynamis Envirotech, Ottowa who claimed damages over the non delivery of the cargo of its sister ship, the *Nadezhda Obukhova*. A High Court Admiralty Division judge in London refused the company's application to make the captain of the *Khudoznik Pakhomov* deliver the cargo. The ship returned to Canada with its cargo. The Canadian government halted further cargoes and cancelled the contract with ReChem.[50]

The controversy over the Canadian PCB shipment was condemned by Environment Secretary Chris Patten: "No-one seriously committed to improving our environment should seek to manipulate and play on public fears and worries by distorting risks and disregarding the value and effectiveness of our monitoring and controlling procedures and discrediting the standards of British technology in waste disposal."[51] Sir Hugh Rossi explained the cause of the government's anger and said the "hysteria" was causing "such unnecessary alarm and damage to an industry which is becoming an important contributor to the nation's balance of payments."[52]

Imports of waste from the north of Ireland have been constant since the beginning of the seventies, with the growth of the chemical industry in the south. In 1981 the British government intervened to prevent Penn Chemicals of Cork exporting its toxic waste for landfill at Chemstar, a Manchester solvent recovery factory, which was forced to close after an explosion killed a worker. Penn said the waste, potassium methyl sulphate, was "smelly but basically non toxic."[53] Experts at Harwell laboratories said little was known about the chemical but that it should be chemically treated or incinerated, never dumped.[54] Penn began exporting the waste to ReChem.[55] In March 1982 the problems of Penn's waste arose again when the *Craigantlet*, a ship bound for Liverpool from Belfast, ran aground off the Scottish coast. Firemen in protective clothing attempted to save the leaking container and prevent spillage into the sea but they were unsure what they were dealing with. The ship's documents stated that the waste was methyl sulphate salts. The Scottish Office later said it was potassium methyl sulphate. Penn claimed it was sodium methyl sulphate.[56]

The port of Briton Ferry is comprised of a number of small wharfs on the banks of the river Neath which opens into the Bristol Channel. No waste had been handled there until 1987. Between June 1987 and September 1989, 30,000 tonnes of waste came through the port.

The first shipment, contaminated Dutch soil, was docked in 1987. The Environmental Health Department were asked to become involved to allay public fears, as it was rumoured that the soil was radioactive. When the soil was sampled, it was found to contain heavy metals and cyanide. It was not, however, classified as special waste.

In Spring 1988 Leigh Environmental used Briton Ferry for the importation of (West) German fly ash. A stockpile of 4,500 tonnes accumulated on the quayside awaiting transportation. The ash contained calcium carbonate and heavy metals but could not be classified as toxic or special waste.

The local authority reviewed their control over the storage and handling of wastes and issued a licence for the handling of non-special wastes to the wharf owner, as a means of controlling their importation. The German ash and the Dutch soil continued to come through the port but public concern died down.

On 30 August 1988 the Environmental Health Department were contacted by a customs officer. He said the *Karin B* intended to enter Bristol Channel and unload its cargo of toxic waste on to lighters for transhipment to a wharf at Briton Ferry. The Environmental Health Department told the customs officer that the wharf was only licensed to receive non-special waste. Eventually Environment Minister Virginia Bottomley said the waste could not land because the waste was not precisely described in its documentation and it wasn't properly labelled.

Public confidence in the council's assurances that no toxic waste was being imported were shattered. In the wake of the *Karin B* affair, speculation over the status of the German ash continued. A local journalist acquired a German TV video of the ash being unloaded at Briton Ferry by dockers not wearing protective clothing and workers at a Walsall waste disposal facility, with full body protection, transferring the waste by air transfer system to silos. The journalist contacted Herr Gruber, a chemist at Leverkusen, where the waste originated. He said the waste was hazardous but his manager disagreed. Public concern intensified with the council contesting claims of toxicity, having analysed the waste themselves.

The vessel *Katja*, sister ship of the *Karin B*, docked quietly on the 13 September 1988 and unloaded bulk red ash from Denmark which was destined for Leigh in Walsall. Notice had not been given to the council or the wharf operator. The wharf operator notified the council and the ash was analysed. The local analyst told the council the waste was "special" because of its arsenic concentration. The council realized licence conditions had been contravened and they were worried that the public would not differentiate between the German fly ash and this new waste. The wharf operator said he had been misled by the importer.

The mound of 3,000 tonnes of red ash had to be made safe because it was drying out and being dispersed by the wind. The fire service covered the waste, wearing full protection and using independent air supplies. The event received much media coverage. Over the months the council pressed Leigh to remove the waste from the quay, but it remained there until Spring 1989 because Walsall council were unhappy about the disposal arrangements.

In 1987 Franco Rafaelli, an Italian businessman who had been living in Nigeria for ten years, wrote to an Italian firm, SI Ecomar. He told them he was willing to import their industrial wastes into Nigeria and dispose of them by burning. Between August 1987 and May 1988, five boatloads of Italian waste went to Koko, a small port on the Niger Delta.

More than 10,000 drums of assorted chemicals were stacked out in the open in the yard of a farmer named Sunday Nana. The drums were exposed to the heat and to tropical storms. They began to deteriorate and leak. A village was drawing water from a nearby well, and Mr Nana used several of the drums to store his own food and water. The drums began to smell and local people fell ill.

The dump was discovered in 1988, when the media were tipped off by Nigerian students in Italy. President Babangida had just returned from an Organisation of African Unity summit, where he had denounced African governments who "collaborated" with foreigners to dump toxic waste in Africa. Rafaelli fled the country without a passport. Nigeria withdrew its ambassador from Italy and seized two Italian ships unconnected with the dumping.

The Italian government agreed to remove the drums of waste from Koko although it had no obligation to do so. They ordered Ambiente, a waste handling subsidiary of the state owned chemical company, ENI, to pick up the waste, bring it back to Italy for disposal and clean up the dump. Ambiente chartered two German ships, the *Karin B* and the *Deepsea Carrier*, to collect the waste. The Italian government wanted the waste repacked properly before loading onto the ships. Three Nigerians fell ill while handling the waste and the Nigerian government insisted it be loaded immediately.

The *Karin B* set sail for Ravenna in Italy. The Mayor and citizens of Ravenna formed a blockade across the port and informed Rome that they would not allow the *Karin B* into the port. It went to Cadiz in Spain, where it was also refused entry. On 30 August 1988 the *Karin B* appeard at Briton Ferry, where it was also forbidden to land its cargo. It arrived at Cherbourg in September. Five of the crew had fallen ill because of the fumes from the drums of waste. The captain requested that the men be evacuated but the French Authorities at Cherbourg refused. A navy doctor was flown out to the ship. The *Karin B* sailed from France but was refused entry in Belgium, Germany, France and Holland. On 2 September 1988, the Italian government ordered the *Karin B* and the *Deepsea Carrier*, which was still at sea, back to Italy. They also banned the export of toxic wastes to developing countries. The wastes from the *Karin B* were unloaded at an Italian military port and stored by a company called Mont Eco at four sites in Northern Italy.

Laura Bosisio of ENI, the company who loaded the *Karin B* wastes, said there were no PCBs among them. Earlier three separate teams in Koko – one from Friends of the Earth, one from the UK Atomic Energy Authority and one from the US Environmental Protection Agency – all found PCBs in the waste. The *Karin B*'s original destination when it left Nigeria was an Italian high temperature incinerator owned by Monedison which is Italy's only large facility capable of dealing with PCBs.

In October 1990, Torfaen Borough Council were notified of a consignment of PCBs coming to ReChem in Pontypool. Allan Woods, the company's spokesman, said he believed the consignment from Italy had originally been part of the cargo of the *Karin B*. A senior official for Torfaen Borough Council said the authority could not legally refuse the cargo provided that it was in safe condition and that ReChem could properly dispose of it. Lanstar Wimpey Waste, in Cadishead, said they had taken more than 100 tonnes of waste from Mont Eco.

In 1987 a Maltese ship, the *Lynx*, sailed from Italy bound for Djibouti in Africa. It was carrying more than two thousand tonnes of resins, pesticides and PCBs. The wastes had been collected from Italian paint and chemical producers by a waste disposal firm called Jelly Wax. When the *Lynx* arrived off Djibouti the port authorities refused to allow it to be unloaded as they suspected the waste was radioactive. The *Lynx* sailed to Venezuela and 11,000 drums of waste were unloaded and stored in the open. After a few months the drums began to leak gasses, which made nearby residents ill, and led to the death of a young boy. The Venezuela government ordered the removal of the drums. A Cypriot ship, the *Makiri*, took the waste to Syria where a waste disposal company was paid £100,000 to dispose of it, but the Syrian authorities ordered the cargo to be removed before disposal could take place. The owners of a rusty freighter, the *Zanoobia*, stepped in and the waste was loaded on to the vessel. The *Zanoobia* sailed to Greece but was turned away.

The *Zanoobia* headed back to Italy after more than a year at sea. At this stage the crew were desperate. Most of them were suffering from skin, kidney and respiratory illnesses and one of them died. The authorities in Italy refused to allow the rusty and badly leaking drums to be offloaded and dockers refused to handle them. The *Zanoobia* had to wait for more than a month for permission to land at Genoa. The Italian government were forced to take responsibility for the *Zanoobia* after Syria claimed it was carrying radioactive waste. The waste had come from a US airbase in Northern Italy where about 800 nuclear warheads were deployed. The ship's captain said he had carried 2,500 drums, some of them were not properly sealed. A barrel caught fire at one stage. He said the crew had no idea how dangerous the cargo was, and were shocked to hear it was radioactive. The port authorities at Genoa called in Italian army specialists to repack the chemicals in new containers.

In 1990 a deal to ship the waste from the *Zanoobia* to the UK was arranged by Castalia, a company in which the Italian government is a major shareholder. In 1988 Leigh Environmental pulled out of a deal with Castalia after press disclosures. The head of Castalia, Robert Ferraris said the British firm had insisted on an escape clause in the contract to allow withdrawal in the event of public outcry.

In September 1990 Wimpey Lanstar admitted processing waste from the *Zanoobia*, but denied taking dangerous toxic waste such as PCBs. John Parker of Wimpey Lanstar said the waste was mostly paint polymers, residues and varnishes. He admitted: "If we had known it was from the *Zanoobia* we wouldn't have accepted it. I suppose we should have asked a few more questions." Wimpey Lanstar had been given a licence by GMWDA to dispose of 271 tonnes of the *Zanoobia* cargo. Chief officer of GMWDA, Alec Davidson, said that waste was a national responsibility which should be dealt with as part of the production process. He said: "It certainly shouldn't be hawked around the world like this. We deplore the whole business but it's enshrined in law and we can't stop it." Erica Woods, of Community Campaign Against Toxic Wastes, said: "We don't want any of it coming here. Why should it be disposed of here in Europe's biggest industrial estate, near factories, near where people are living."

A senior Italian government official, Admiral Guiseppe Francese, was responsible for disposing of the waste from the *Zanoobia*. Reacting to the report that the waste was being shipped to Manchester, he told the Genoa daily *Il Lavoro*, "It's not that Italy doesn't have appropriate incinerators, but I prefer to send [the waste] elsewhere, to avoid protest demonstrations by environmentalists".

The transfer of the *Zanoobia* waste to Cadishead provoked strong community opposition. In October 450 protesters marched to Wimpey Lanstar and handed in a 1,000-name petition demanding the end of the treatment of the *Zanoobia* waste. General manager Nigel Dibben said: "We're not in the business of causing alarm, and if people are disturbed about it then their feelings must come before hard commercial factors." Lanstar said they would receive no more waste from the *Zanoobia*.

Notes and references

1. HMIP, First Annual Report 1987–88.
2. Environment Committee, 2nd Report, *Toxic Waste*, 1988–89.
3. Ibid.
4. HWI, Third Report, 1988.
5. Environment Committee, 1988–89.
6. HMIP, Provisional data on transfrontier shipments.
7. HMIP, *Transfrontier Shipments Figures*, Summary Total Report, 1 January 1989.
8. Hewitt, M.R., Shanks and McEwan, "The Transfrontier Directive and the Principles of Importation of Waste into the UK", 1988. Symposium, "Importation of Waste to the UK", 8 September 1988. See also GMWDA "Waste Disposal Draft Plan", p 72.
9. *Yorkshire Post*, 15 June 1988.
10. Hewitt, M.R., 1988.
11. *Hull Daily Mail*, 2 February 1990.
12. Hewitt, M.R., 1988.
13. HMIP, *Transfrontier Shipments Figures*.
14. *Hull Daily Mail*, 9 November 1987.
15. *Yorkshire Post*, 15 June 1988.
16. *Hull Daily Mail*, 2 February 1990.
17. Taylor, Brian; interview with Robert Allen 1991.
18. *Hull Daily Mail*, 9 November 1990.

19. Notes of meeting held at County Hall, Beverly, 30 November 1988.
20. HMIP, Foreign Imports through Felixstowe, 1/ 4/ 89–23 February 1989.
21. *Daily Times*, 21 March 1990.
22. *Yorkshire Post*, 15 June 1988.
23. Meeting of TGWU, Paul Redgate, George Douse and Fred Beech, 22 March 1990.
24. *Daily Times*, 21 March 1990.
25. Greenpeace Policy Statement, "The International Waste Trade to the UK".
26. Hewitt, M.R., 1988.
27. *Yorkshire Post*, 15 June 1988.
28. Hewitt, M.R., 1988.
29. *New Statesman*, 18 August 1989.
30. HWI, June 1988, Third Report.
31. Hewitt, M.R., 1988.
32. Valette, Jim, Greenpeace International, Waste Trade Inventory, 1988.
33. Environment Committee, 2nd Report, Toxic Waste, 1988–89.
34. Environment Committe, Press Release, 10 Aug 1989.
35. Environment Committee, 1988–89.
36. *Independent*, 19 September 1990.
37. HWI, Third Report, June 1988.
38. Ibid.
39. *Independent*, 19 September 1990.
40. In an interview with *She* magazine, 1990.
41. *Digger*, 28 Jan 1988.
42. Organisation for Economic Co-operation and Development, *The State of the Environment* (Paris: OECD, 1991).
43. Powell, David; letter to Donald McDonald, 6 September 1988.
44. Martel, Holly; Officer of the Minister of the Environment, Canada; letter to D. Powell, 31 January 1989.
45. Powell, David; letter to Donald McDonald, 31 July 1989.
46. McLellan, Ronnie; Canadian High Commission; letter to David Powell, 7 August 1989.
47. Gouvernement du Quebec, Communiqué, 11 August 1989.
48. *New Statesman*, 18 August 1989.
49. *Daily Post*, 10 August 1989.
50. *Guardian*, 25 August 1989.
51. Department of the Environment, *Statement on the Movement and Disposal of Toxic Waste*, 14 August 1989.
52. *New Statesman*, 18 August 1989.
53. *New Scientist*, 3 December 1981.
54. Ibid.
55. Ibid.
56. *New Scientist*, 11 March 1982.

Table 7.1: Quantities of waste imported via named ports

PORT	1986–87	1987–88	1988–89
Avonmouth	1,519	2,737	–
Barking	–	2,276	–
Boston	–	1,283	164
Bristol	1,333	–	–
Briton Ferry	2,300	7,536	12,360
Chatham	1,730	49	88
Dartford	4,122	4,133	52
Devonport	25	–	–
Dover	1,638	3,559	20
Felixstowe	13,038	21,416	12,311
Fishguard	50	–	–
Fleetwood	1,956	2,173	1,440
Garston	203	–	–
Greenwich	–	2,276	–
Grimsby	271	14	–
Harwich	855	1,578	141
Holyhead	124	384	221
Immingham	12,233	293	20,020
Ipswich	543	15	290
Liverpool	80	658	1,405
Manchester	60	–	–
Newhaven	23	48	–
Newport	3,082	–	315
Plymouth	42	–	–
Poole	230	72	–
Portsmouth	70	17	–
Purfleet	100	109	83
Rainham	924	–	–
Ramsgate	25	–	395
Seaforth	127	–	–
Seaham	3,599	–	–
Southampton	13	222	–
Teesport	–	1,650	2,959
Tilbury	2,666	2,910	41

Source: Written Answers to Parliamentary Questions (30 November 1989).

Table 7.2: Toxic waste imported into Heysham (1989)

Inorganic acids	1.5
Cyanoacrylate residue	39.8
Dry cake (welding waste)	17.5
Fluegas dust	0.0
Glycolmethacrylate scrap	156.4
Laboratory wastes	2.6
Lead waste	182.7
Lime tungstate sludge	38.4
Mercury waste	0.6
Organic waste	16.3
Paracetamol waste	15.3
PCBs	34.0
Petroleum wastes	3.7
Pharmaceutical waste	11.0
Plating wastes (mixed)	3.4
Sodium hydroxide	4.0
Stripping compounds	7.0
Sulphuric acid	2.8
Zinc hydroxide sludge	12.8
Irish	516.3
Non-Irish	33.4
Total	549.7

Table 7.3: Toxic waste imported into Scotland (1987 and 1988)

Year	Amount (tonnes)	Country of origin	Destination (District Council)	Nature of waste
1987	15	Holland	Strathkelvin	sub-soil
1988	2	Sweden	Inverness	Sodium metal
	451	Ireland	Edinburgh	Solvent and contaminated methylated spirit (for recycling)

Information is ...

Source: S...

...able for earlier years.

Chapter 8

COMPANIES AND SITES

Community opposition to toxic waste disposal is primarily based on the argument that the companies are insensitive about the effects on human and animal health and on the environment. Methods of disposal, whether they be treatment, recovery, recycling, landfill or incineration, are not, the communities argue, as safe as the waste industry would like to believe they are. Wastes which are defined hazardous or special (also known as toxic) are not always landfilled and/or incinerated separately, but are frequently mixed with other wastes before or during disposal. It is this method of waste management that communities are worried about. "We don't know what is going in and we don't know what is coming out. Of course we fear for our health and our children's health," is a frequent complaint. "Fear of the unknown," is another often-heard cry. HMIP, the principal regulating authority for air emissions, has believed since its inception that the end-products (emissions from incineration) of toxic waste disposal are correctly monitored and are within British and EC statutory limits. The waste disposal industry says the same, yet it is worth noting the attitude of the National Association of Waste Disposal Contractors when they welcomed the Environmental Protection Bill of 1990. "Any control system is only as good as its enforcement and unless the Government ensures that the right control authorities with adequate resources are available the new measures will not work."[1] The communities opposed to toxic development have known, for most of the eighties, that "enforcement" is merely a word in a nebulous vocabulary and that "adequate resources" are not and have not been available to regulating authorities ... ... industry is also aware that govern- ... ces ... ot and cannot adequately ... ies. The waste disposal ... ... have no trust in self ... ment and the regulating authorities do n ... ... ement industry monitor the industry and that the communi ... ... laint. regulation. Yet the waste disposal or waste ... will tell you it cannot understand why there is so muc ...

"The difficultly is credibility," argued Jim Caldwell of Motherwell Bridge. "There is so much going on at the present moment in the waste business. The more people become aware the more they start arguing about it."[2] "HMIP is grossly understaffed," said Peter Bateman of Rentokil. "I think there is enough legislation there, what it does need is to be properly policed. Obviously responsible companies who have got a reputation to look after are very concerned to do the right thing, but we all know what there are cowboys and people doing unauthorized tipping and so forth who would bring the waste disposal (industry) into disrepute, whereas industry as a whole is probably trying very hard to do the right thing."[3] Ross Ancell, managing director of Wimpey Waste Management, argued that "done properly, landfill or incineration are the best that is available". The new bill, he said, would raise standards and drive out the cowboys. "It has to be done properly. The days of people just opening up an old fashioned dump or tip and disposing of anything in a willy-nilly manner are gone. Companies such as our own have the engineering skills and the resources to design sites that are fully contained that can control leachate and control gas. We see ourselves as a waste management company, not a waste disposal company."[4]

There is no doubt that the majority of landfill sites in Britain are contaminated today because the legislation did not exist 20 years ago to prohibit indiscriminate toxic dumping by "cowboys", or anyone else for that matter.[5] Yet the community opposition of the eighties and nineties is not principally against the "cowboys" or the small operators, but is against large companies such as Cleanaway, Leigh, Lanstar, ReChem, Wimpey Waste. The so-called cowboys may have been responsible for contaminated landfills or for smoke and smells from antiquated boilers but they are not, communities argue, the major problem. It is convenient for individual companies and understandable that they should defend their own activities and seek to blame others. However, many of the small waste disposal operators of 20 years ago are today directors of the major operators. And many of the smaller operators were once subsidiaries of larger, seemingly respectable, companies. For example ReChem were once a subsidiary of British Electric Traction (BET). The operators of two of Britain's five high-temperature incinerators in 1980 were small companies bought out by larger companies. The change of ownership changed little for the communities.

Britain is now in its second wave of waste disposal expansion. This time the expansion is much greater, so much that if all the proposals received planning permission there would be excess capacity.[6] The first wave brought merchant incinerators to Ellesmere Port, Hucknall,

Roughmute, Pontypool and Fawley. The second wave has brought expansion to Ellesmere Port and Fawley and new proposals for high capacity merchant, sewage and clinical incinerators in Derry (by Du Pont in league with the Irish and British authorities), in Doncaster (by Leigh), Trafford Park (by Leigh), Seal Sands (by Ocean Environmental, a subsidiary of Cory Waste), Teeside and Tyneside (by International Technology Corporation in league with Northumbrian Water Authority), Humberside (by ITC), Scotstoun (by Motherwell Bridge), Kelso (by John Salvesen and Blue Circle International Waste Management) and smaller applications from local authorities and health and water boards.

During the eighties three companies offered liquid incineration services in Inverness (Nontox, later Lanstar), in Rye (Gelpke & Bate) and in Birmingham and Newcastle (Robinson Bros) to clients with small amounts of waste to get rid of. Between them they offered a capacity of approximately 6,500 tonnes. In 1984 ReChem closed their high-temperature incinerator in Roughmute and in 1987 Berridge Incinerators in Hucknall sold their plant to Leigh in Killamarsh. In 1988 merchant incineration capacity in Britain was approximately 82,000 tonnes (see Table 8.1, p. 192). Leigh did not appeal the decision by the local authority to turn down their applications for Trafford Park, but they did appeal against the decision in Doncaster. Caird also appealed against the decision by the authorities in Glasgow. Although Motherwell Bridge Products have stated categorically that they would not burn hazardous or toxic waste, the community regarded their proposal with suspicion.

By the end of the eighties the toxic waste disposal business was booming. Leigh returned record profits in March 1989; turnover was £51 million, profits £4 million. Cleanaway also returned a record profit for their operations in 1988, £9 million on a turnover of £70 million. And ReChem increased their profits from £2.64 million in 1988 to £5.15 million in 1989. Turnover increased from £13.38 million to £19.47 million. Although the increases in profit and turnover did not continue to accelerate as dramatically in 1990 (ReChem for example only made £5.71 million on a turnover of £21.06 million) it was clear to the private sector that there was plenty of money to be made out of toxic waste. The property firm, Caird, had decided in July 1987 that the environmental services sector was so profitable that they acquired 13 waste disposal and environmental services companies in the same year and bought a 5.4 per cent holding in Leigh. Caird also attempted to buy out waste contractors Wistech in a £7.5 million deal. Caird pulled out after an investigation of Wistech's accounts. The return on Caird's new investment was swift. The first six months in the waste business brought a pre-tax profit of £466,280. For the six month period to December 1988 the pre-tax profit was £1.7 million on a turnover of £5.7 million.[7]

With ReChem and Cleanaway dominant in the incineration market throughout the eighties it was inevitable that someone would challenge their monopoly. Leigh has been amongst the fastest growing companies in the industry. A product of the industrial revolution, Leigh developed from a freight company on Britain's canals to become a dominant force in the specialized waste and treatment sector by the end of the 1980s. At the time Leigh bought the Berridge's incinerators in 1987 for its Killamarsh operation, the company was able to boast that it could offer its customers chemical treatment, solvent recovery, oil recovery, deep mine disposal and incineration at its sites in Birmingham, Runcorn, Sheffield and Southampton. In 1989 Leigh treated an estimated 400,000 tonnes of toxic waste. Its operation in Birmingham is, boast the company, the largest waste treatment and disposal site in Britain. Leigh say that 200,000 tonnes of liquid wastes can be stored at Killamarsh.[8]

When Caird, ITC, Leigh and Ocean Environmental submitted their applications for new incinerators, the Scottish waste disposal company, Shanks and McEwan, dominated landfill disposal. In 1986 Shanks and McEwan shifted 1.4 million tonnes of waste. By 1991 they were involved in treatment, liquids, sludges and drummed waste – in total, 7 million tonnes of waste. In 1986 Shanks and McEwan had acquired London Brick's landfill operation from Hanson Trust to become Britain's dominant landfill company. In December 1990, after a close look at the waste disposal business and its projected growth, Shanks and McEwan moved into incineration with the purchase of ReChem. The enlarged Shanks and McEwan group now has 8 per cent of the waste disposal market. Chief executive Roger Hewitt said, a couple of months after the ReChem take-over, that the rate of growth in the waste disposal would continue "for a long time. It is not a business much likely to be affected by recession".[9] Hewitt also said he believed Shanks and McEwan could take 15 per cent of the market and perhaps 25 per cent. "We still have got a lot of money to be made in the UK, a lot of growth."[10]

Before the Shanks and McEwan acquisition of ReChem, the London Stock Exchange acknowledged the performance of the six quoted waste disposal companies. This included these two companies plus Attwoods, Caird, Leigh and H.T. Hughes. All gave impressive performances on the stock market in 1988 and 1989. The waste disposal market in Britain is believed to be worth £1 billion a year (£1,000 million in Scotland). Yet the industry is still fragmented, with more than 4,000 active companies.[11] While there was a spate of acquistions of the smaller firms throughout the seventies and eighties it is still the large companies who are vulnerable to take-overs. Before Shanks and McEwan acquired ReChem in December 1990, Leigh announced an agreed bid for H.T. Hughes while Severn Trent Water made a bid for Caird. Observers noted the Severn Trent move with interest because the privatized water authorities, notably Severn Trent, Northumbrian, Southern, Thames, Welsh and Yorkshire, who want to bury sewage sludge, have set up subsidiaries and see the waste market as a natural extension of their business. Although 75 per cent of Attwoods business is in the US, the British waste disposal market has not been ignored by US companies. Waste Management and Browning-Ferris Industries, both US companies, are very active in Europe and are looking to Britain.[12]

There are many reasons for the growth of the waste disposal industry in Britain – the continued growth of waste generated each year and the salient fact that government sees no real problem with its importation. In 1989 the House of Commons Environment Committee's second report on Toxic Waste reported that 2,505 million tonnes of waste had been generated in England and Wales. Of these wastes 4.5 million were classified as hazardous and special.[13] The Hazardous Waste Inspectorate reported that the total waste produced in Scotland for 1989 was 8.67 million tonnes. Special waste totalled 71,000 tonnes.[14] In Ireland the amount of waste produced is unknown and figures for hazardous and/or dangerous waste vary. Consultants to the Irish government estimated that approximately 4,000 tonnes of toxic waste were exported in 1989, to Britain, France and Finland. HMIP stated that in 1989 1,357 tonnes of Irish waste were imported through seven ports in England and Wales. The consultants believe that between 3,000 and 6,000 tonnes of hazardous waste are produced each year, yet between 1985 and 1988 figures provided by government and industry estimated hazardous waste produced in Ireland at between 75,000, and 20,000 tonnes.[15] In the north of Ireland 2.26 million tonnes of waste were generated in 1988, of which 8,289 tonnes were special. During 1988 an estimated 330 tonnes of special waste were sent to England for treatment or high temperature incineration.[16] Approximately seven hundred tonnes of toxic waste are sent by Du Pont from Derry to France and Finland for high temperature incineration each year.[17] There are approximately 5,000 waste disposal sites in England and Wales[18], in Scotland there are 835.[19]

Notes and References

1. Hamer, Mick "Long arm of the law", *Municipal Journal*, 2 February 1990.
2. Caldwell, Jim. Interview with Fiona Sinclair, 1990.
3. Bateman, Peter. Interview with Fiona Sinclair, 1990.
4. Ancell, Ross. Interview with Fiona Sinclair, 1990.
5. In May 1990 Friends of the Earth in conjunction with the *Observer* launched a Toxic Tips campaign. A year later FoE demanded that the government release information on the precise location of over a 1,000 potentially dangerous, gassy landfill sites in England and Wales. FoE's Sean Humber said there needed "to be a thorough national survey of all landfill sites, both open and closed, to assess the extent of the risk". Government, he added, should make the original polluter pay. For further information on the FoE's Toxic Tips campaign contact FoE: 26–28 Underwood Street, London N1 7JQ. See also FoE press releases, 10 May 1990 and 4 April 1991 and briefing document "Sitting on a pollution time bomb".
6. In *Haznews* 12, March 1989 ReChem managing director Malcolm Lee said there was a great risk of over-capacity. Nick Crick, marketing manager of Cleanway, commented that wastes currently landfilled by some producers could go to incinerators if the capacity was available. Communities and environmental groups fear this will encourage the companies to import foreign waste.
7. Company annual records and accounts 1988, 1989, 1990. See also *Haznews*, February 1988–present. Park House, 140 Battersea Park Road, London SW11 4NB, and ENDS, Unit 24, Finsbury Business Centre, 40 Bowling Green Lane, London EC1R 0NE.
8. Leigh company records and accounts, 1989, 1990.
9. *Scotland on Sunday*, 27 January 1991.
10. Ibid.
11. *Financial Times* survey "Waste Management", 26 September 1990.
12. Ibid.
13. House of Commons Environment Committee, 2nd Report, Toxic Waste, vols 1–3, March 1989 (HMSO, 1989).
14. Hazardous Waste Inspectorate Scotland, report 1987–1990, HMSO
15. Allen, R. and Jones, T. *Guests of the Nation*, pp 296–298 (London: Earthscan, 1990). See also Byrne O'Cleirigh, "The incineration of hazardous waste in Ireland – a feasibility study", *Irish Times*, 12 October 1988 (DoE figures on toxic waste generated); *Irish Independent*, 8 October 1988; Rick Bolens of the IIRS, Confederation of Irish Industry and DoE quoted in *Cork Examiner* 16 September 1987.
16. Aspinwall & Co, "A Review of Waste Disposal in Northern Ireland", June 1990.
17. Du Pont, "The facts", press release, January 26, 1991.
18. House of Commons, Environment Committee, 2nd Report, Toxic Waste.
19. Survey conducted by the Scottish Development Agency (now Scottish Enterprise) in 1989.

Table 8.1: Merchant incineration in Britain (1988)

COMPANY/LOCATION	CAPACITY (tonnes)	Type of waste
ReChem: Fawley	15,000	Liquids
ReChem: Pontypool	32,000	Solids
Cleanaway: Ellesmere Port	20,000	Liquids
Leigh: Killamarsh	8,700	Liquids
Nontox/Lanstar: Inverness	3,000	Liquids
Robinson Bros: Birmingham	1,000 (estimate)	Liquids
Robinson Bros: Newcastle	1,000 (estimate)	Liquids
Gelpke & Bate: Rye	1,400 (estimate)	Liquids
Approximate total:	82,000	

Table 8.2: Merchant capacity in Britain (projected)

Company/location	Capacity[1] (tonnes)
ReChem: Fawley	40,000
ReChem: Pontypool	32,000
Cleanaway: Ellesmere Port	48,000[2]
Leigh: Killamarsh	13,100[3]
Robinson Bros: Birmingham	1,000 (est)
Robinson Bros: Newcastle	1,000 (est)
Gelpke & Bate: Rye	1,400 (est)
Approximate total:	140,000 tonnes

Notes

1. If the proposals by Blue Circle (in Kelso), Caird (in Renfrew), Du Pont (in Derry), ITC (in Humberside, Teeside and Tyneside) Lothian Chemical Co. (in Edinburgh), ICI and BP (in Grangemouth) and Ocean Environmental (in Teeside) were successful the capacity would rise to an estimated figure of 400,000 tonnes. And this does not include proposals by companies for clinical sewage incinerators. At the end of 1991 there were still 12 proposals for high temperature incinerators in Britain and Ireland. This would lead to excess capacity. Not enough hazardous waste would be generated in Britain and Ireland to meet this capacity.

2. When Cleanaway began to operate their second incinerator in Ellesmere Port their original model was still in operation, which gave them a total capacity of 68,000 tonnes.

3. Leigh operate two incinerators in Killamarsh with 8,700 and 4,400 tonnes capacity.

Sources: Company literature, planning documents, *Haznews* February 1988 – present

Table 8.3: Total waste (England and Wales)

Type of waste		Quantity (million tonnes)
Liquid industrial effluent		2,000
Agricultural		250
Mines and quarries (inc. china clay)		130
Industrial		50
Hazardous & special	3.9	
Special	1.5	
Domestic and trade		28
Sewage sludge		24
Power station ash		14
Blast furnace slag		6
Building		3
Medical wastes (clinical)		0.15
Total		2,505.15

Source: House of Commons Environment Committee 2nd
    Report Toxic Waste

Table 8.4: Value of landfill sites (1990)

Company	Capacity (m cu m)	Value (£m)
Shanks & McEwan	80	220–280
Leigh	50	100–125
Caird	60	90–120
Attwoods	30	60–90
HT Hughes	15	34–45

Source: *Financial Times* 24 September 1990/Citicorp ''The UK
    Waste Management Industry''

Table 8.5: Landfill sites in Scotland

District/Islands council	Landfill sites		Other waste diposal facilities
	Open	Closed	
Berwickshire	3	0	0
Ettrick and Lauderdale	5	2	0
Roxburgh	3	0	0
Tweeddale	1	1	0
Clackmannan	7	3	1
Falkirk	19	5	9
Stirling	8	0	1
Annandale and Eskdale	3	3	2
Nithsdale	6	0	1
Stewartry	10	2	0
Wigtown	9	0	1
Dunfermline	19	6	2
Kirkcaldy	17	6	1
North East Fife	10	2	4
Aberdeen	16	4	6
Banff and Buchan	19	10	6
Gordon	12	4	1
Kincardine and Deeside	9	7	0
Moray	9	2	5
Badenoch and Strathspey	2	2	0
Caithness	7	1	1
Inverness	1	1	2
Lochaber	8	2	0
Nairn	0	0	0
Ross and Cromarty	5	3	0
Skye and Lochalsch	3	0	0
Sutherland	8	0	3
East Lothian	10	6	1
Edinburgh	13	15	7
Midlothian	9	12	0
West Lothian	19	14	3
Argyll and Bute	1	0	2
Bearsden and Milngavie	0	0	1
Clydebank	1	3	1
Clydesdale	8	3	0

Table 8.5 (continued)

District/Islands council	Landfill sites		Other waste diposal facilities
	Open	Closed	
Cumbernauld and Kilsyth	9	6	1
Cumnock and Doon Valley	3	3	1
Cunninghame	16	7	8
Dumbarton	5	1	4
East Kilbride	10	1	2
Eastwood	3	1	2
Glasgow	17	24	16
Hamilton	10	0	3
Inverclyde	4	4	4
Kilmarnock and Loudoun	4	3	2
Kyle and Carrick	4	3	1
Monklands	6	6	1
Motherwell	8	3	7
Renfrew	17	15	3
Strathkelvin	13	5	1
Angus	11	6	0
Dundee	6	3	4
Perth and Kinross	24	8	4
Orkney	15	4	2
Shetland	6	0	3
Western Isles	7	0	5
Total	478	222	135

Source: Written Answers to Parliamentary Questions (24 April 1990).

Chapter 9

DIOXINS, FURANS, TCDD AND 2,4,5-T

There is no safe level for dioxins, the environmentalists argue. Industrialists say there is, and refer to the hypothesis that elements of polychlorinated dibenzo-para-dioxins and polychlorinated dibenzofurans, commonly known as dioxins and furans, have always been in the environment. Some scientists agree that the burning of wood for heating and cooking and other naturally occuring combustion processes have generated dioxins and furans. This hypothesis has been shattered in recent years with the revelations, based on data derived from the analysis of ancient human tissue, that the source of dioxins and furans is recent and anthropogenic.[1]

This debate has been a refrain of the early nineties but what scientists are sure about is that dioxins and other related chemical compounds are lethal to certain animals. Dioxin is regarded as the most powerful carcinogen ever tested in laboratory animals. A millionth of a gram of the most studied and most toxic dioxin known (2,3,7,8-TCDD) will kill a guinea pig.[2] Where scientists disagree is the effect of dioxins and furans on humans; there isn't, some insist, enough evidence to show any serious long-term effects on human beings.[3] A 1981 review of the literature on dioxin by a panel of the American Medical Association did not find "any documented human deaths, directly attributed to dioxin poisoning".[4] More recently, the US EPA claimed that studies of dioxin on people yielded inconsistent results.[5] Yet, as long ago as 1977, the International Agency for Research on Cancer (IARC) identified the toxic effects of TCDD to include dermatological disorders (notably chloracne), liver damage and disease, heart disease, kidney and lung disorders, neurological damage and depression.[6] Dr Alastair Hay has said that from his research and from the evidence available "dioxin has the ability to cause cancer itself, rather than being merely a cancer promoter".[7] What is certain is that scientists need to learn more about human exposure to dioxin. In the 1989 government report 'Dioxins in the Environment', the Committee on Toxicity said: "We recommend that the most useful action which could be taken, is as far as possible to identify the remaining major sources of these chemicals and to take appropriate measures to reduce inputs to food, consumer products and the environment with the aim of reducing human exposures."[8]

The major sources of the 75 polychlorinated dibenzo-para-dioxins known to exist, particularly TCDD, are chemical manufacture and incineration. (Exhaust fumes from cars running on leaded petrol and paper produced in a chlorine bleach also contribute dioxins and furans to the environment.) Dioxin emissions from these sources may be dispersed by wind and water. They become present in the soil, may be ingested by grazing livestock, and lodge in their fatty tissue and milk. In this way they enter the food chain. Humans may also be exposed to dioxins and furans through inhalation, by drinking contaminated water and through dermal contact with contaminated products. "The recognition that everyone is potentially exposed to dioxins and furans through contact with soil or certain consumer products, through food or through breathing air has raised the inevitable question – are the levels to which people are exposed likely to be harmful?"[9] According to Tony Gatrell, who co-authored *Burning Questions: Incineration of wastes and implications for Human Health* with Dr Andrew Lovett in 1989, "inadequate data and lack of information about exposure" is a major problem in Britain.[10]

Communities and environmentalists are angry that information about the health effects of chemical compounds like dioxins and furans has not been gathered and assessed before now. Evidence of the health effects is limited to the results of industrial accidents (see Table 10.1 p.212), the effects of "Agent Orange" in Vietnam, exposure to herbicide spraying, laboratory experiments on animals and a US government-sponsored study of 5,172 chemical workers from 12 factories, who had occupational exposure to TCDD.[11] In general terms dioxin toxicity in humans can be placed in two categories: short, high exposure such as industrial accidents and long, low exposure such as in Vietnam. Workers in factories manufacturing the chemical compounds which create dioxins and workers handling the phenoxy herbicides, 2,4-D and 2,4,5-T (which are contaminated by TCDD) also fall into the second category.

Most scientists now believe that dioxin may cause chloracne, headaches, nausea, itching, liver and kidney damage, heart and lung disorders, neurological damage and neuropsychiatric symptoms.[12]

In the long term dioxin is also believed to affect the reproductive and immune systems. Recent studies have conclusively linked dioxin with cancer.[13] Dioxins are known to lodge in fatty tissue; scientists in the USA recently announced they could detect dioxin that had persisted in a human body for decades with the aid of new technology which allowed them to trace the chemical in blood tests.[14] It can take about seven years for half the dioxin in our bodies to be excreted, thus the poison can be released into the body if there is a sudden loss of fat at any time. Dioxins are metabolized in the liver, which releases enzymes to break them down, but the same enzymes damage the liver itself. The thymus gland in babies is particularly vulnerable to dioxin damage and as babies depend on the thymus for immunity they get immunosuppression. Sterility may also be an effect of dioxin poisoning. Some workers have complained of loss of libido, insomnia and irritability following exposure to TCDD.[15]

Industrial accidents, such as that at Monsanto in 1949 and at Seveso in 1976 (see Table 10.1, p. 212), have exposed over two thousand workers to high levels of TCDD and many of these workers (and in the Seveso accident, residents) developed chloracne, among other illnesses. TCDD is also a by-product of the manufacture of 2,4,5-trichlorophenol, which is used in the manufacture of herbicides, notably 2,4,5-T, an ingredient with 2,4-D of Agent Orange, which the US used during the Vietnam war to clear forest and undergrowth. Although TCDD has been known as a contaminant of 2,4,5-T for more than 30 years and that exposure to TCDD (in whatever form) causes chloracne it has only been in the past three years that conclusive evidence has linked dioxin to cancer.

In 1989, while the US study on chemical workers was being completed, a (West) German epidemiologist claimed he had established the first evidence to link dioxin poisoning with cancer. On 17 November 1953 an explosion at a BASF trichlorophenol factory in Ludwigshafen exposed between 122 and 153 workers to TCDD. In September 1989 Dr Friedemann Rohleder produced a report, at the 9th international conference on dioxin, in Toronto, Canada, in which he recorded an unexpectedly high incidence of cancer in the BASF workers. Eight workers, he claimed, had died from cancer.[16]

The BASF explosion has been extensively studied. Before Rohleder was invited to do his study nine known papers had chronicled the details of the explosion and the effects of the toxin on the workers. The earlier studies reported that 53 of the workers who had been exposed to TCDD had developed chloracne, 42 severely; 21 of these suffered consequent damage to internal organs or disturbances of the nervous system, and a prominent feature was liver damage. The son of the one of the workers developed chloracne following contact with his father's clothes. In 1958 another case of poisoning was identified in a 57-year-old worker involved in repair work on the contaminated site. He died nine months later, and chloracne was evident. The autopsy report showed he had died from degeneration and inflammation of the pancreas, and the coroner attributed the cause of death to exposure to the chemical. In a later study of the 53 workers, 22 were still working at the factory, 16 had retired and 15 had died. Of the 22 workers still employed in 1976 four suffered long-term effects; two had persistent chloracne (23 years after the accident), one had paralysis of the left leg and one had permanent loss of hearing. The 16 men who had retired were well. Of the 15 deaths two had committed suicide and four had cancer, at 54, 64, 66 and 70 years of age. In 1982 when Alastair Hay published an account of the accident he recorded that 17 workers had died. Six, he said, were due to cancer, "four of which involved the gastrointestinal tract". These figures, he claimed, were highly significant because they represented twice the expected rate for any of the control groups. "More significant," he wrote,

> is the fact that two of the gastrointestinal cancers occurred in workers between the ages of 65 and 69. Compared with age-matched controls the recorded deaths are 10 times higher than expected. A second epidemiological study of this group of men conducted three years later comes to the same conclusion. Commenting in the incidence of stomach cancers in the dioxin-exposed workers, the authors say that it was ". . . .considerably greater than expected and cannot be adequately explained as mere chance event".

The three epidemiological studies and Rohleder's subsequent evaluation of the BASF accident were the first to reveal that workers exposed to dioxin may develop cancer.[17]

Rohleder came to the BASF case after an investigation of the mortality records of the exposed workers, published in 1985 on behalf of the Born Berufsgenossenschaft (BG) – the industry association which handles liability claims, dismissed associations between the exposure to TCDD and cancer. As a result of the BG report compensation was refused; the victims and their families decided to take legal action. The German government stepped in and invited Rohleder to evaluate the evidence. Rohleder examined BASF's medical files. The data for the original BASF study, which the BG used to compile their 1985 report, showed that 153 workers had been exposed. The most recent set of data, which the BG compiled because of the controversy over the compensation claims, showed that 122 workers had been exposed. Rohleder compared both sets of data and excluded 20 supervisors, who had been included by BASF in the original records, as Rohleder believed they had not been exposed to the TCDD. The suspicion was, he said, that someone had tried to minimize the observable effects of the dioxin exposure. He also excluded two other workers. He was left with what was a micro-epidemiological study of the available medical records and subsequent evidence. He concluded: "This analysis adds further evidence to an association between dioxin exposure and human malignancy," and that occupational studies, especially the historical type "should not be regarded as valid, unless there is company-independent confirmation of their completeness and correctness". He also added: "The true mortality and especially the morbidity experience of persons who were highly exposed to dioxins following the BASF accident, still remains to be studied with adequate methods."[18]

Some scientists believe TCDD is a potent cancer promoter. Hay, how-ever, is convinced that dioxin can *directly* cause cancer. Along with the BASF analyses the US study, which was published in January 1991 and which conclusively reported that TCDD is a carcinogen, supports this belief. "Mortality was evaluated in a large group of chemical workers with exposure to phenoxy herbicides and chlorophenols contaminated with TCDD and with sufficient latency for the expression of some occupationally related cancers," the US report noted and suggested that the estimates of effect in the study might provide an upper level of risk to be anticipated in humans. The report concluded that there was a 15 per cent excess of all cancers but that, for several types of cancer previously associated with TCDD, excesses were not observed. The exception was soft tissue sarcoma, "for which a nine-fold increase was seen among workers with more than one year of exposure and 20 years of latency".[19]

Closer to home, an industrial accident involving TCDD over 20 years ago is still shrouded in secrecy. An hour or so after midnight on 23 April 1968 an explosion at the Coalite and Chemical trichlorophenol works in Bolsover, Derbyshire, showered workers with TCDD and killed the duty chemist. Since 1965 the fine chemicals unit of the company had manufactured 2,4,5-trichlorophenol (TCP) in the pro-duction of the herbicide 2,4,5-T. Within a month of the accident and until 8 December 1968, 79 cases of chloracne were recorded, many severe. The majority of workers recovered within four to six months. Three years after the explosion two outside contractors were exposed to a contaminated tank, despite rigorous cleaning of plant equip-ment after the accident, and developed chloracne. The son of one worker and the wife of another also developed chloracne some months later.[20]

In 1976, three years after the Bolsover and Coalite GP, Dr George May, had reported the health effects of exposure in the workers, he told a local newspaper there had been other chemical leakages from other factories which had also caused chloracne. May claimed these accidents had been covered up. He alleged that one of the leakages had led to the closure of the factory for several months. He claimed the effects of chloracne were more severe than amongst the Coalite workers.[21] It was around this time that the company became less willing to release information.

Shortly after the accident at Seveso, the HSE asked the company to re-evaluate the health of its workforce. George May took on the task in consultation with the HSE. Three outside consultants were also commissioned to do part of the research, Drs Eric Blank and Anthony Ward from Sheffield University and Dr Jenny Martin from Chesterfield Royal hospital. Alastair Hay takes up the story:

> Coalite informed Martin some time after she had completed her work that it did not wish to have the information published. She was also informed of the nature of the control group used for the study. Realizing that the study had been devalued by Coalite's choosing to include management staff in the control population instead of restricting it to chemical workers alone – the recognized practice in this type of study – Martin arranged a second study.

The second study (blood chemistry from eight Coalite workers suffering from chloracne was compared with a matched control group) was carried out without Coalite's involvement and Martin subsequently published the results in *The Lancet* in February 1979. Shortly after this Martin's home was broken into and the medical records of the eight workers stolen. Nothing else was taken. She had no duplicates.

Coalite, when it released an abbreviated version of Martin's first study, stated that there were no differences in the blood parameters between the dioxin-exposed and control group. Martin, in her full report to the company, said there were. Furthermore the HSE had been unable to obtain the full medical records from Coalite. The HSE therefore accepted Coalite's abbreviated medical report, which did not reveal the real health of the workers.

The following year Hay revealed in *Nature* magazine that Coalite would not agree to further studies of its workers if the results were submitted for publication. The publicity prompted the HSE to request the full records from Coalite. When these were handed over the HSE's Employment Medical Advisory (EMAS) recommended that the workers exposed to dioxins and still employed by the company should partake in medical tests every 3 to 5 years. When George May attempted to repeat the tests a few months later he was only able to examine 29 of the 41 workers who had participated in the 1977/78 survey. Critics said this sample could not be expected to yield meaningful results.

Coalite's persistent secrecy throughout meant that the majority of workers exposed to dioxin did not get the opportunity to undergo medical tests. Coalite's 1977/78 survey included only 41 of the 90 workers exposed. The sponsor of this study said it could not support further studies if Coalite insisted that the results could not be published. Hay commented:

> The repercussions of the Coalite affair are likely to continue for some time, and the company is not alone in having laid itself open to criticism. The actions of the HSE over Coalite have been brought into question by ASTMS, the union most involved with the situation. It is the union's view that the Executive was caught out by Coalite. And to compound the problem the union claims that the HSE failed to keep it informed of developments at the firm, as it had promised. For the HSE, Coalite has been a chastening experience and it is unlikely to be caught so unprepared in future. The developments at Coalite have shown that the legislation to secure a safer working environment in Britain is still far from perfect and it is clear that some action is necessasry to remedy this situation. For the dioxin-exposed workers at Coalite the remedy they need is assurance that their health will not be impaired in the long term."[22]

In the 1989 DoE dioxin report the authors refer to current "epidemiological studies of former employees of a trichlorophenol plant" who have had symptoms of chloracne. If the DoE is in fact referring to Coalite it is interesting to note that the Medical Division of the HSE is not, despite the DoE reference, involved in the study. However, the HSE's epidemiological department in Bootle said a report on the Coalite incident was being prepared and was expected in 1992.[23]

Scientists, toxicologists and epidemiologists in Britain agree that, compared with other countries, information on dioxin and furans is limited. It is debatable whether Britain (particularly the authorities) has learned anything from the experiences of other nations, notably Germany and the US – where the dioxin controversy has raged for much of the past three decades. In the US the use of 2,4,5-T in the late sixties during the Vietnam war and in herbicide spraying became publicly known, yet it wasn't until the eighties that the US governmental agencies began to take serious measures to protect people from dioxin poisoning. In Britain the controversy over 2,4,5-T was played down by government; dioxin, it said, wasn't a major problem.

In 1983 scientists at a Centre for Disease Control meeting agreed that a safe level for dioxins in soil was one part per billion.[24] The following year the US EPA announced that it would extend the regulations on hazardous waste to include all 75 dioxins and 135 furans by March 1985. Anything used to produce energy or make a fuel, such as spent solvents, would be regulated.[25] In 1987 the EPA proposed that the unit risk of dioxin exposure should be reduced to 0.1 picograms per kilogram of body weight per day (0.1 pg/kg bw/day). The current level was 0.006 picograms or a trillionth of a gram per kg of body weight, the amount of dioxin which, if ingested daily for 70 years, would result in cancer in one in a million.[26]

When the Women's Environmental Network (WEN) in Britain released their dioxin briefing document it stated that the guidelines for levels had been exceeded in some areas and among some people, according to preliminary results from a government survey. "For example a child weighing 20 kg drinking one pint of milk a day from farms in rural areas would receive 1.28 pg/kgbw/dayfrom milk alone, not including butter, cheese, meat and fish, and 7.13 pg/kgbw/dayif drinking milk from an urban area or near an incinerator." WEN is concerned that children are not being taken into account in studies (notably the MAFF study of foods for dioxins and furans and the DoE dioxin report).

> A child weighing one third of the adult weight of 60 kg will always have three times the intake if s/heeats or drinks the same amount of food. . . . Children eat less than adults, although they drink more milk. The New Zealand government acted when a child consuming a litre of milk a day would receive a significant proportion of 5 pg/kg bw/day. Children in [Britain] will exceed this amount from a pint of milk per day from industrial areas or near incinerators. Children are growing rapidly so a higher intake may be diluted as their body weight rises. A child's immune system is very dependent on the thymus gland which is a target organ for TCDD.[27]

In the DoE dioxin report the authors state: "Once dioxins are present in food, no subsequent treatment can remove them. Levels in food can only be reduced affectively by controlling or eliminating emissions of dioxins into the environment."[28]

WEN's concern is the extreme persistence of dioxins, in soil, in humans: "Once exposed, we continue to receive low doses from our body fat, thus increasing the chance that some cell-level accident will occur eventually. No animal has been studied for the equivalent of a human lifetime. Since there is a real increase in cancers in older people which is not explained by an increase in the number of people surviving to older age, it may be possible that we are seeing the long-term effects of a reduction in immunity caused by dioxin-like substances in the body for perhaps 40 years."[29]

Notes and references

1. The theory that dioxins and furans are the result of natural combustion has been quoted extensively from the Department of the Environment (1989) document, "Dioxins in the Environment" (Pollution paper no. 27, HMSO; see pp. 11, 12 & 51). The contrary argument is discussed in several papers but see "Sources of Dioxins in the Environment: Second Stage Study of PCDD/Fs in ancient human tissue and environmental samples", Tong, Gross, Schecter, Monson and Dekin; *Chemosphere* Vol 20, nos 7–9, pp. 987–992 (1990).
2. Women's Environmental Network (WEN), draft paper on health effects of dioxins (1991).
3. Tschirley, F.H. (1986) "Dioxin", *Scientific American*, 254 (2), pp. 21-27.
4. *New York Times*, 4 July 1983 7.
5. Dowd, R., "EPA revisits dioxin risks", *Environ. Sci. Technol.* Vol 22, no 4, 373 (1989).
5. *Some fumigants, the herbicides 2,4-D and 2,4,5-T, chlorinated dibenzodioxins and miscellaneous industrial chemicals* (IARC, 1977). See also *Long term hazards of polychlorinated dibenzodioxins and polychlorinated dibenzofurans* (IARC, June 1978). IARC internal technical report No 78/0001.
7. Hay, A. *Science*, Vol 221, 1162.
8. DoE (1989), 80.
9. Ibid, 50.
10. Gatrell, T. correspondence with Robert Allen, 1991.
11. Fingerhut, M.A. et al, *Mortality among US workers employed in the production of chemicals contaminated with TCDD*. NTIS PB 91-125971 (US Department of Health and Human Services (CDC).
12. WEN (1991).
13. See Fingerhut (1991) and Rohleder, F. (1989) "Dioxins and cancer mortality: Reanalysis of the BASF cohort". Presentation to the 9th International Dioxin symposium, Toronto, Canada, September 17–22, 1989.
14. *New York Times*, 14 October 1986.
15. Kimborough, Renate D., *Occupational exposure in Halogenated biphenyls, terphenyls, napthalenes, dibenzodioxins and related products* (Netherlands: Elsevier, 1980) pp.373–397.

thalenes, dibenzodioxins and related products (Netherlands: Elsevier, 1980) pp.373–397.

16. Rohleder (1989).
17. See Hofmann (1957), Goldmann (1972, 1973) refs on pp. 000 and Theiss, A.N. and Frentzel-Beybe, R., "Mortality study of persons exposed to dioxin after accident which occurred in the BASF on 13 November 1953", 5th International Medichem Congress, San Francisco, September 5-9 1977; IARC (June 1978); Theiss, A.M., Frentzel-Beyme, R and Link, R. "Mortality study of persons exposed to dioxin in a trichlorophenol-process accident which occurred in the BASF AG on November 17, 1953". *Am. J. Ind. Med.* pp. 179-189 (1982); Hay, A., *The Chemical Scythe*, (New York: Plenum Press, 1982) 102–105 and Lehnert and Szadkowski *Arb. Soz. Prav. Med* 20. 225–232 (1985).
18. Rohleder (1989). See also Dioxin: A Briefing from WEN, 3; and "New analysis links dioxin to cancer", *New Scientist*, Oct 28 1989.
19. Fingerhut (1991)
20. May, G. (1973) "Chloracne from the accidental production of TCDD", *Brit. J. Ind. Med.* 30, pp. 276-283 (1973).
21. *Derbyshire Times*, 13 August 1976; *New Scientist*, 19 August 1976.
22. Hay, A. "Chemical company suppresses dioxin report", *Nature*, 284, p. 2 (1980); "Coalite health survey talks", *Nature*, 285, p. 4 (1980); Dioxin hazards: secrecy at Coalite. *Nature*, 290, p. 729 (1981). See also Hay, A. *The Chemical Scythe*, pp. 109–121 (New York: Plenum Press, 1982).
23. HSE, interview with Robert Allen.
24. *New York Times*, 1 July 1983.
25. *New York Times*, 22 December 1984.
26. The Committee on Toxicity, in the DoE dioxin report state: "We have derived conservative estimates of levels above which there may be concern about adverse effects in humans and these are 60 pg/kg bw/day, 1 pg/kg bw/day and 10 pg/kg bw/day for immunotoxic, reproductive and carcinogenic effects respectively. Because of the considerable uncertainties inherent in deriving these figures, these levels should not be regarded as "acceptable daily intakes" or "tolerable daily intakes". Instead we recommend that the lowest figure identified, 1 pg/kg bw/day, be considered as a guideline value in the sense used by the WHO, that is a level which, when exceeded, should trigger investigation and appropriate measures to reduce environmental levels generally." WEN state: "There is no level known to be safe, because studies are based on animal experiments and on effects of industrial accidents where other toxic substances are usually involved as well. The US EPA has a 'virtually safe level' based on a level of risk of one in a million; i.e. 0.0064 pg/kgbw/day."
27. WEN, "Dioxin. A Briefing" WEN.
28. DoE, xi.
29. WEN (1991).

Chapter 10

TOXIC WASTE AND HEALTH

> Epidemiology is a discipline that seeks to describe and account for variations in ill health. Few attempts have been made to gauge the extent to which badly managed waste disposal may contribute to health problems.[1]

The objectives of an epidemiological study centre on identifying causes and their effects, by linking exposure to a source to the potential or actual health effects. For the sake of this argument the source would be a waste disposal facility, such as an incinerator, a landfill or a storage depot for chemicals. This process may seem relatively clear cut, but it involves complex, expensive and wide ranging research, and careful consideration of statistics and methodology. A study must be designed and conducted so that meaningful and significant results are yielded; no two situations are the same. However, every study will engage with the problem of how to separate the source in question from background factors, which could also be causing health problems.

An epidemiological study involves four basic phases: 1. Hazard identification – indentification of and quantification of the toxic chemicals that may be present; 2. Dose response analysis – determination of the kinds of effect that may be caused by the chemicals that are present, and analysis of how the effects may be expected to vary in frequency and severity with the level of exposure; 3. Exposure assessment – assessment of the extent to which human beings are or may be, exposed to the chemicals that are present; and 4. Risk characterization – estimation of the numbers of persons who are likely to be affected as a result of exposure and characterization of the types and severity of toxic effects that are likely to be experienced by a person who is afffected.[2] The data collected during this process will be correlated, and patterns may become apparent.

Toxic waste disposal involves a diverse range of substances which are destroyed or dumped in many different ways. The nature of this process presents difficulties in identifying which toxins are present, as chemicals at waste disposal facilities are usually encountered in mixtures. This is particularly true in the case of incineration of wastes, as new compounds such as dioxins and furans may be formed during the cooling process. Liquids may leach into soil from landfill becoming solids or sludges. "It is important to realise that conducting even a simple survey of health effects is perilous if one lacks information about toxins and exposure, since few – if any – measurable health effects are sufficiently specific for exposure to a particular toxin or groups of toxins to be surrogates for directly measuring toxic exposure," Clark W. Heath, Jr of the Chronic Diseases Division of the Centers for Disease Control stated.[3]

It is necessary to determine which chemicals are present and in what amounts. This process may involve soil sampling, air monitoring, and tests on water and foliage. The nature of the chemicals must be considered. Some chemicals such as polychlorinated biphenyls (PCB) are persistent – this quality which makes them so useful to industry makes them extremely toxic (they are not broken down so they accumulate). The difficulty is isolating the toxins, as many chemicals may be present. In the case of airborne emissions, chemicals may combine with backround pollution to give higher concentrations.

When the chemicals have been identified their toxicity must be assessed. "For the bulk of the chemicals in commercial production, the available data are insufficient to permit even the most rudimentary toxicological evaluation."[4] Evaluation of the toxicity may, in a lot of cases, rely on experimental work done on laboratory animals. However, the value of toxicity testing on animals is questionable, as different species will exhibit different effects. Much of our knowledge of the effects of toxic chemicals on humans stems from industrial accidents.

It is necessary to evaluate the relationship between the level of exposure to a chemical or chemical mixture, and the frequency and severity of the toxic reaction that may result or be expected to result from the exposure.[5] The degree of risk to individuals from various exposure levels can then be estimated, but this process is, again, hampered by the lack of information about the effects of toxins on humans. In a clinical drug trial a cross-section of healthy individuals who are similar in age and build are used. They are given increasing doses of the drug and blood samples are monitored at regular intervals. Adverse reactions in relation to dose level can be assessed and compared with the control group (a group of people who are given a placebo or inactive compound). Assessing exposure and effect of chemicals found on disposal sites could be compared to conducting a clinical trial using unknown amounts of several chemicals on a diverse group of individuals.

Chemicals found on disposal sites are usually combined, so they may interact with each other, and the effects of a combination will be different from the effects of individual chemicals. "With a few exceptions, such as the combined effect of sulphur dioxide and particulates, there is insufficient information at present to establish guidelines for mixtures."[6] Moreover, individual chemicals may produce many different effects. WHO noted: "It is exceedingly unusual for a particular biologic effect to be the specific and exclusive result of a particular toxic exposure."[7] Illnesses such as cancer can have a long latency period between exposure and tumour development. The long-term follow-up required is expensive and needs considerable commitment. "Although thresholds are known to exist for most types of toxic reactions, no threshold is known or presumed to exist for mutagenic effects, carcinogenic effects and teratogenic effects, at least so far as their induction by certain toxic agents is concerned," noted Heath.[8]

Background factors complicate an investigation, and information must be gathered about these other exposures; for example, whether an individual is exposed to smoking, alcohol, drugs or to other chemicals at work. Details such as sex, age and socio-economic status are useful as individuals may be at risk to lesser or greater degrees. Children will be more vulnerable to the effects of toxins as they have a smaller body mass. Breastfeeding babies may ingest chemicals through mothers' milk. Elderly people may suffer from chronic illnesses which could be aggravated by exposure to toxins. Individuals with asthma and chronic respiratory diseases will be affected to a greater degree by airborne pollution. Those in the lower socio-economic group may have health problems related to poor nutrition and bad housing. "Thus the population must be considered very heterogenous in respect of response to air pollutants. Existing information does not allow adequate assessment of the proportion of the population that has an enhanced response."[9]

To estimate the number of people likely to be affected by exposure to a source of toxins the information gathered must be integrated, but this process is likely to be hampered by uncertainty and lack of data. "Because of the length, complexity and cost of each step in the process, as well as the inherent uncertainties, a detailed and comprehensive attempt to arrive at a risk characterisation, through the entire process has not been attempted for any disposable site, to our knowledge.[10]

The underlying statistical problem in epidemiology is in estimating the risk of a specific disease among people living near a source of pollution and comparing it with the risk to the population in general.[11] A "cluster" is seen as "any group of cases that is sufficiently large and spatially concentrated as to be unlikely as to have arisen by chance".[12] Cases could be individuals with a specific illness, such as cancer. Methods are required to establish whether such clusters are statistically significant. There may be reason to believe that "the cases are related to each other by a social or biological connection, or that they may have a commmon relationship with some other event or circumstance."[13]

Gatrell and Lovett[14] applied a method which relates disease incidence to distance from a possible source of pollution. They conducted a study to assess whether there was any connection between the distribution of laryngeal cancers and proximity to the site of a former industrial waste incinerator in Lancashire. Data on cancers between 1974 and 1983 were used. Mapping was done to detect visually any clustering of cases near the plant. It became apparent that the distribution of most cancers reflected the distribution for the population as a whole, with the exception of laryngeal cancers.

In a further report Gatrell and Lovett stated: "Supplementary analyses have confirmed the earlier findings, namely that the incidence of cancer of the larynx in part of Lancashire is associated with proximity to the site of a former industrial waste incinerator."[15] However they do not claim that the former incinerator actually caused the cancers. "Clearly, in the absence of additionl information it is foolish to claim that living near the incinerator has 'caused' cancer of larynx."[16] The latency period for cancer development will vary for different individuals. They had no information on other risk factors for individuals, such as smoking histories or alcohol consumption. There was no background environmental information on air pollution.

Congenital eye malformations have been noted in the areas around merchant incinerators in Britain. The eye defects included anopthalmia (absence of an eye) and micropthalmia (tiny eyes). This defect normally affects one baby in a 10 milllion population per year. Richard Collin, a consultant eye surgeon at Moorfields Eye Hospital in London said: "The incidence in these two areas makes one extremely concerned." Four children were born in the Pontypool area between 1980 and 1984 with these defects: one child was born with tiny eyes, two born with only one eye and one was stillborn with no eyes. Five babies with eye defects were born within 13 months of each other at Bonnybridge. Around the same time twelve calves were born on one farm with deformed or missing eyes and a one-eyed kitten was born.[17]

The Welsh and Scottish Offices conducted studies into the incidence of malformations and found no evidence of a raised incidence of eye malformations in the areas around the ReChem plants. The Welsh Office report stated: "There were no cases notified in the 1975–1984 or 1976–1985 period of babies from Torfaen with the conditions of micropthalmos or anopthalmos."[18]

The Welsh Office did find a significant increase in the incidence of babies born to Torfaen mothers with ancephalus (absence of part of the brain and skull) and with polydactaly (extra fingers or toes). "The analysis for the 1975–1984 and 1976–1985 periods confirms that the incidence of notified cases of babies born to Torfaen mothers with ancephalus or with polydactaly are significantly higher than would be expected by chance given the rates prevailing in Gwent and Wales as a whole." The report comments on the introduction of ultrasound screening for neural tube defects such as spina bifida and ancephalus. The screening allows these defects to be detected early in pregnancy and the pregnancy may be terminated. This factor would have reduced the number of defects notified. Dr Anthony Jones, a consultant radiologist specializing in the ultrasound detection of foetal abnormalities in a Welsh hospital, commented on the Welsh Office statistics. He said that many defects are detected early in pregnancy and are subsequently terminated so would not be included in the statistics. "Statistics concerning babies born with congenital abnormalities are not necessarily indicative of the true incidence of the abnormality unless the statistics take into account efforts made to detect abnormality early in the pregnancy thereby allowing early termination of affected pregnancies." Jones suggested that the male–female ratio be examined, as the alteration of the male–female ratio is one of the indications of chemical pollution.[19]

Gatrell and Lovett have set in motion the first national study of eye malformations in children, in collaboration with consultant surgeons and opthomologists at Moorfields Eye Hospital, London. "We cannot fail to reach the conclusion that some previous work has relied on anecdote and speculation and that we need to bring to bear to such problems the combined skills of epidemiologists, geographers, and statisticians in researching this important area of public concern."[20]

Notes and references

1. Gatrell, A.C. and A.A. Lovett, 1986, "Hazardous Waste Disposal in England and Wales", *Area*, 18, 275-83.
2. Upton, Arthur C., Theodore Kneil *et al*, 1989, Public Health Aspects of Toxic Chemical Disposal Sites.
3. Heath, Clark W. Jr., 1988, Field Epidemiological Studies of Populations Exposed to Toxic Waste Dumps.
4. National Academy of Science, National Research Council, 1984, Toxicity Testing: Strategies to Determine Needs and Priorities.
5. Upton, Arthur C. et al, 1989.
6. World Health Organisation, 1987, *Air Quality Guidelines for Europe*.
7. Office of Science and Technology, 1985, *Chemical Carcinogens: A Review of the Science and its Associated Principles*.
8. Heath, Clark W. Jr., 1988.
9. World Health Organisation, 1987.
10. Upton, Arthur C., Theodore Kneip et al, 1989.
11. Hills, Michael and F. Alexander, 1989, *Statistical methods used in assessing the risk of disease near a source of possible environmental pollution: A review*.
12. Gatrell, A.C., Research Report No. 9, On Modelling Spatial Point Patterns in Epidemiology: Cancer of the Larynx in Lancashire.
13. Knox, E.G., 1988, Detection of cluster, in P. Elliot (ed.) *Methodology of Enquiries into Disease Clustering*. See also: Gatrell, A.C., 1990, A spatial statistical model of the incidence of cancer of the larynx in North West England, North West Regional Laboratory, Research Report, to appear. Cuzick, J. and R. Edwards, 1990, Spatial Clustering for Inhomogenous Populations.
14. Gatrell, A.C. and Andrew A. Lovett, Research Report No. 8, *Burning Questions: Incineration of Wastes and Implications for Human Health*. Diggle, P.J., A.C. Gatrell and A.A. Lovett, 1990, Modelling the incidence of cancer of the larynx in part of Lancashire: a new methodology for spatial epidemiology, in R.W. Thomas (ed.) *Spatial Epidemiology*, forthcoming.
15. Gatrell, A.C., Research Report No.9.
16. Gatrell, A.C. and A.A. Lovett, Research Report No.8.
17. "The Cyclops Children", *New Society*, 17 Jan 1983, pp. 104-5.
18. Welsh Office, 1987, The Incidence of Congenital Malformations in Wales, with particular reference to the district of Torfaen, Gwent. An updated analysis of notifications to the Office of Population Censuses and Surveys.
19. Jones, Anthony Dr., 1984, Letter to Nicholas Edwards.
20. Gatrell, A.C. and A.A. Lovett, Research Report No. 8.

Table 10.1: Industrial accidents involving toxins

Company	Place	Year	Exposure
Monsanto	West Virginia, USA	1949	228 cases included chloracne, melanosis, muscular aches and pain, fatigue, nervousness and intolerance to cold at TCP factory. (Firestone 1977: IARC 1978)
—	Germany	1952	60 chloracne cases in two TCP factories (W. Hergt cited by Bauer et al 1961)
Badische Anilin und Soda Fabrik Ludwigshafen, (BASF)	Germany	1953	53 chloracne cases at TCP factory (Hofmann 1957: Goldman 1972, 1973)
—	Hamburg, Germany	1954	31 chloracne cases at 2,4,5–T factory (Bauer et al 1961; Kimmig & Schulz 1957; Schulz 1957)
Rhone–Poulenc	Grenoble, France	1956	17 chloracne cases at TCP factory (Dugois & Colomb 1956, 1957; Dugois et al 1958)
—	Italy	1962	5 chloracne cases at TCP factory (Hofmann & Meneghini 1962)
Phillips Duphar	Amsterdam, Holland	1963	Of 50 people affected, 44 developed chloracne, 10 still had skin complaints in 1976 and 4 workers died within 2 years of the accident. According to Hay some 30–200g TCDD were released; the factory was sealed off for 10 years, then dismantled, embedded in concrete and dumped at a deep point in the Atlantic: 2,4,5–T factory (Dalderup 1974; Hay 1976)
Dow	Midland, Mich, USA	1964	60 chloracne cases at 2,4,5–T factory (Firestone 1977)
Diamond Alkali Co	Newark, New Jersey, USA	1964	29 chloracne cases at 2,4–D & 2, 4, 5–T factory (Bleiberg et al 1964; Firestone 1977)

Table 10.1 (continued)

Company	Place	Year	Exposure
Rhone–Poulenc	Grenoble, France	1966	21 chloracne cases at TCP factory (Dugois et al 1968)
Spolana	Czechoslovakia	1965–69	76 chloracne cases following exposure to TCDD between 1965 and 1969 at 2,4,5–T & PCP factory (Jirasek et al 1973, 1974)
Coalite	Belsover, Britain	1968	79 chloracne cases at TCP factory (Milnes 1971; Jensen & Walker 1972; May 1973)
—	USSR	1964–70	128 workers showed skin lesions and of 83 examined, 69 had acne as a result of occupational exposure at 2,4,5–T factory (telegina & Bikbulatova 1970)
JCMESA	Seveso, Italy	1976	An accident at the TCP factory resulted in the contamination of a densely populated area (four towns, approximately 100,000 people). Five days after the accident animals began to die. Almost 20 days after the accident 700 people were evacuated from the heavily contaminated areas. Hundreds complained of acute skin lesions and symptoms of systemic poisoning. By end October 1976 37 cases of chloracne had been reported among the evacuated, mostly in children and young people. More than 500 residents were treated for toxic symptoms and 134 chloracne cases have now been confirmed. The malformation rate rose from 1.03 per 1000 births in 1976, to 13.7 in 1977 and 19.0 in 1978. A cancer registry has been established but Hay has said that any impact on cancer rates will not be revealed for many years because of the long latency

Table 10.1 (continued)

Company	Place	Year	Exposure
JCMESA	Seveso, Italy	1976	period between exposure and tumor development. Under-reporting before and after the disaster was also apparent so figures may not be comprehensive. (Bert et al 1976; Hay 1976; Hay 1982)

Sources: American Medical Association
Hay, A. (1979) Accidents in trichlorophenol plants: a need for realistic surveys of a certain risk to health.
Vol 15 Lyon: IARC 1977.

Industrial accidents have exposed more than 2000 workers to large amounts of TCDD. (The figure is imprecise because of incomplete data on several accidents and details of occupational exposure.)

TCDD: 2,3,7,8-Tetrachlorodibenzo-para-dioxin
2,4,5-T: 2,4,5-trichlorophenoxyacetic acid (herbicide)
PCP: Pentachloropenol
2,4-D: 2,4- dichlorophenosyacetic acid (herbicide)
TCP: 2,4,5-trichlorophenol

Toxic effects in humans exposed to TCDD have occurred after; the occupational exposure during the production of 2,4,5-trichlorophenol (TCP) and 2,4,5-T; exposure in factories and in the surrounding environment following accidents during the production of TCP and 2,4,5-T; and exposure to herbicides and other materials containing TCDD.

TCDD is formed during the production of 2,4,5-trichlorophenol (TCP). Because TCP is the precursor of the herbicide 2,4,5-T, TCDD is often produced during its manufacture. TCDD has been identified as a component of the products of explosions involving the production of TCP and 2,4,5-T. As TCDD is a contaminant of 2,4,5-T, exposure during the spraying of the herbicide has resulted in toxic symptoms. "Agent Orange" used during the Vietnam war, was a 50:50 mixture of 2,4-D and 2,4,5-T which contained up to 30 mg/kg or more TCDD.

References

Firestone, D., "The 2,3,7,8–TCDD problem: a review". In Ramel C., ed., *Chlorinated Phenoxy acids and their dioxins: Mode of Action, Health risks and Environmental Effects. Ecol. Bull.* (Stockholm), 27 (1977).

IARC, *Polychlorinated and polybrominated biphenyls. Monographs on the evaluation of the carcinogenic risk to chemicals of humans.* (Lyon: International Agency for Research on Cancer, 1978).

Bauer, H., Schulz, K.H. & Spiegelberg, U. Berufliche Vergiftungen bei der Herstellung von Chlorphenol–Verbindungen. Arch. Gewerbepath. Gewerbehyg. 18
538–555 (1961).

Hofmann, H.T., "Neuere Erfahrungen mit hoch-toxischen Chlorkohlenwasserstoffen" Naunyn–Schmiedeberg's *Arch. exp. Path. Pharmakol.*, 232, 228–230 (1957).

Goldman, P.J., "Schwerste akute Chlorakne durch Trichlorphenol- Zersetzungprodukte" *Arbeitsmed. Sozialmed. Arbeitshyg.*, 7, 12–18 (1972).

Goldman, P.J., "Schwerste akute Chlorakne, eine Massenintoxikation durch 2,3,6,7– TCDD" *Der Hautartz*, 24, 149–152.

Kimmig, J. and Schulz, K.H., "Berufliche Akne (sog. Chlorakne) durch chlorierte aromatische zyklische" *Ather. Dermatologia*, 115, 540–546 (1957).

Dugois, P. and Colomb, L., "Remarques sur l'acné chlorique (à propos d'une éclosion de cas provoqués par la prèparation du 2,4,5–trichlorophénol)" *J. Med. Lyon*, 38, 899–903 (1957).

Dugois, P., Marechal, J. and Columb, L., "Acné chlorique au 2,4,5–trichlorophénol" *Arch. Mal. prof.*, 19, 626–627 (1958).

Hofmann, M.F. and Meneghini, C.L., "A proposito delle follicolosi da idrocarburi clorosostituiti (acne clorica)" *G. Ital. Derm.*, 103, 427–450 (1962).

Dalderup, L.M., "Safety measures for taking down buildings contaminated with toxic material" *II. T. soc. Geneesk.*, 52, 616–523 (1974).

Hay, A., "Toxic cloud over Seveso" *Nature*, 262, 636–638 (1976).

Bleiberg, J., Wallen, M., Brodkin, R. and Applebaum, I.L., "Industrially acquired porphyria" *Arch Dermatol.*, 89, 793–797 (1964).

Dugois, P., Amblard, P., Aimard, M. and Deshors, G., "Acné chlorique collective et accidentelle d'un type nouveau" *Bull. Soc. franc. Derm. Syph.*, 75, 260–261 (1968).

Jirasek, L., Kalensky, J., Kubeck, K., "Acne chlorina and porphyria cutanea tarda during the manufacture of herbicides" *Cs. Dermatol.*, 48, 306–317 (1973).

Jirasek, L., Kalensky, J., Kubeck, K., Pazderova, J. and Lucas, E., "Acne chlorina, porphyria cutanea tarda and other manifestations of general intoxication during the manufacture of herbicides" *II. Cs. Dermatol.*, 49,

145–157 (1974).

Milnes, M.H., "Formation of 2,3,7,8–TCDD by thermal decomposition of sodium 2,4,5–trichlorophenate" *Nature*, 232, 395–396 (1971).

Jensen, N.E. and Walker, A.E., "Chloracne: three cases" *Proc. roy. Soc. Med.*, 65, 678–688 (1972).

May, G., "Chloracne from the accidental production of TCDD" *Brit. J. industrial Med.*, 30, 276–283 (1973).

Telegina, K.A. and Bikulatova, L.I., "Affection of the follicular apparatus of the skin in workers occupied in production of butyl ether of 2,4,5–TCD" *Vestn. Dermatol. Venerol.*, 44, 35–39 (1970).

Bert, G., Manacorda, P.M. and Terracini, B., "I controlli sanitari: la sanita incontrollata" *Sapere*, 796, 50–60 (1976).

Hay, A., *The Chemical Scythe: Lessons of 2,4,5–T and Dioxin* (New York: Plenum, 1982).

Conclusion

Wednesday 26 June 1991 dawned with little to offer for sports fans in Britain. Wimbledon would be washed out again and it didn't look good for the third test-match between England and the West Indies. However, it would be a day to remember in South Wales, and it began with a small news story in the *Argus*: Richard Biffa and Malcolm Lee had resigned from ReChem and had given up their executive positions on the board of Shanks and McEwan.[1] The national newspapers did not record the event but three papers, the *Financial Times*, the *Guardian* and the *Scotsman*, noted that Oxford University Press had published the British Medical Association's report on hazardous waste and human health. The *Guardian* reported that the BMA had called for a ban on toxic waste imports. The *Financial Times* and the *Scotsman* reported that the BMA wanted a national waste strategy. There was little evidence that hazardous waste had caused public health problems in Britain, the chairman of the BMA's science and education board had told a press conference, but he added: "We cannot reassure the public there is no risk from these substances."[2] Tara Lamont, who wrote the report, said she was disappointed with the media coverage.[3] The subject, it seemed, wasn't high on the media's schedules. Though how the majority of the national media managed to miss the poisoned milk story in Derbyshire will probably remain a mystery in itself. In the House of the Commons on 26 June, Agriculture Minister John Gummer replied to a parliamentary question from West Lancashire MP Kenneth Hind.[4] The Ministry of Agriculture, Fisheries and Food (MAFF) had announced its findings from a survey of dioxin levels in milk. Gummer explained:

> Sophisticated techniques which have been developed recently in the Department's food science laboratory at Norwich have for the last year made it possible to detect dioxins at minute levels. As milk tends to

show up airborne contaminants, it is a particularly useful and sensitive product to test in a programme designed to protect the public. Dioxins are of course widespread in the environment, and as could be expected, traces of them have been found in all the samples tested.[5]

Gummer added that the scientific and medical experts in his department and in the Department of Health had calculated a maximum tolerable concentration of 0.7 nanogrammes per kilogramme of milk, if the tolerable daily intake (TDI) for dioxin of 0.01 nanogrammes per kilogramme body weight per day, recommended by the EC in 1990, is not to be exceeded. "Their advice," said Gummer on instruction from his department:

is that action should be taken to prevent the direct consumption of milk if it contains a higher concentration of dioxins.

Samples have been taken at three stages; at retail outlets, at the dairies and on the farm. At the retail stage no samples contained more than the normal background levels of dioxin and there is no risk to consumers. At the dairy stage the same applies: this provides further reassurance. Similarly most of the samples taken on farms do not reveal anything unusual. However, on two farms in the Bolsover area of Derbyshire the levels were greater than the threshold set by this Department and by the Department of Health. An extended survey has however indicated lower levels on all the other dairy farms in the vicinity. I am making the results available in the Library of the House. As to these two farms, milk sold is bulked with other milk in the milk tanker and at the dairy. Tests on milk from the dairy have shown background levels. Nevertheless it is best that milk from these farms not be mixed with other milk. My officials have . . . visited the farmers concerned and we have told them and the Milk Marketing Board (MMB) about these results. The MMB has concluded that the milk does not meet the conditions in its standard terms of sale for producers and it will not accept the milk into the food supply.[6]

Gummer concluded: "This department will continue to carry out surveillance in the area and will undertake further detailed studies to learn more about the mechanism for the transmission of dioxins."[7] The news that high levels of dioxin contamination had resulted in the MMB ban on the two Bolsover farms did not excite the national media but it did scare the National Farmers Union (NFU) and the Women's Environmental Network (WEN), and it annoyed shadow food and agriculture minister, Dr David Clark, who criticized the government over its delay in taking action and the secrecy over the results. The government had not indicated the source of the contamination, Clark noted.[8] The NFU, WEN and the environmental movement believed it was localized industrial pollution.[9] WEN, following their own extrapolation of the figures provided by Gummer, concluded that a child who drank a pint of the contaminated milk would exceed the World Health Organization's guidelines for dioxin intake by a count of 20.[10]

Despite the news in May that HMIP had ordered ReChem in Pontypool to improve its operation, the communities did not believe the authorities would be able to take the appropriate action against offenders. The BMA report had offered a glimmer of hope, but the problem of ill-health remained. Studies and reports which had not been peer-reviewed by the scientific community had, like the anecdotal evidence, been excluded from the professional debate, though the BMA noted that "they are often useful in highlighting potential problems which should then be subject to further investigation using proper research protocols". The existing, peer-reviewed, studies had not revealed sufficient data and there was, said the BMA, "a need for more research into the effects of incineration and other exposure to waste on local populations". Yet it was also clear to the BMA that in the absence of reliable studies "it is difficult to make valid risk assessments – particularly regarding the chronic effects of exposure to hazardous substances"[11] The BMA had begun their two-year investigation after Walsall GPs had expressed their concern about ill-health among their patients, during the BMA's 1989 conference. "The BMA, as doctors," said Tara Lamont, "are stating the concerns of the public. We have a mandate, if you like, to state the problem and to give possible solutions"[12]

Almost two weeks after the Derbyshire dioxin results were announced and after the publication of the BMA report, Labour published their proposals for a single environmental protection agency. On the same day, Prime Minister John Major told a conference in London that the HMIP, the National Rivers Authority and the Drinking Water Authority would be merged, replacing a system government said could not be changed because of bureaucratic problems – a system less than four years old. Labour claimed that the government had poached its environment programme. "Labour has been publishing proposals for a single environment protection executive since 1989 . . . Mr Major is just grasping at ideas without any concrete proposals on how to implement them," Labour environment spokesperson Ann Taylor said before the launch.[13]

British politicians have had no problems with the environment. For most of the 20th century successive governments have worried little about pollution. Jon Tinker, in 1972, said Britain's environmental policy was not suited to an advanced industrial society. Offenders "were taken quietly on one side by the prefects and ticked off for letting the side down. There is no need for prosecution; the shame of being found out is reckoned to be punishment enough."[14]

Over the past few decades, very few of Britain's waste disposal companies have been prosecuted. Leigh, with several fines against them, are an exception. John McCormick notes that Britain's "non-coercive approach to pollution control is reflected in the fact that the Alkali Inspectorate prosecuted just three cases between 1920 and 1967. Polluters are regarded as innocent until proven guilty".[15] It's not much better in Ireland. The chemical industry is noted for its pollution, yet only three companies have been successfully prosecuted for pollution offences.[16]

Britain, however, did make a good start. In 1863 Britain created the world's first government environmental agency when it established the Alkali Inspectorate. The job of the inspectorate was to regulate the emission of acid fumes from the alkali industry, which made sodium carbonate for the manufacture of soap, glass and textiles. However, it has only been since Britain's involvement in the EC that new environmental legislation has been put on the statute books. EC legislation, notably on air and water pollution, and toxic waste has forced Britain to introduce tougher environmental laws. Yet McCormick notes:

> Nowhere is the unwillingness of the government to use coercion better exemplified than in pollution control. In countries like the United States and Germany, pollution policy is based on setting agreed standards and targets. In Britain the government has traditionally relied almost entirely on encouraging industry to comply voluntarily with "decent" standards of behaviour. Pollution control laws tend to be broad and discretionary and the regulatory agencies are usually given wide scope to establish and enforce environmental objectives."[17]

Almost twenty years have passed since toxic waste disposal became an issue among communities concerned for their health and for the environment, yet the problem has still not been adequately addressed. Succcessive governments have not dealt with it and the green movement were, until the late-eighties, unable to offer anything of substantial value to local groups.

Some would say that it has been easier for the greens to save the odd seal or whale or dolphin than take on the staid grey-suited bureaucrats of Whitehall or the power of the waste disposal industry. Some green organizations have shown that they are politically and socially naive, unprofessional in research, confidentiality and legislation (particularly in the laws of libel) and in many ways very "green" about the ways of the capitalist world. The whales are a safer option, particularly as it has meant tackling foreign governments with the tacit approval of the home government. Many of those in the environmental movement are insensitive to the issues of employment and unemployment. Saving the world and getting paid for it is one of the newer professions. Sadly, the social and racial mix of Britain is not reflected in the membership of the greens.

Communities opposed to toxic waste disposal in Britain and Ireland have received support and guidance from Greenpeace yet the immediate aims of the communities are often incompatible with the global aims of the environmental group. Fiona Sinclair made the point, following a meeting addressed by a Greenpeace speaker, that the community wanted to know how to deal with the problem of the incinerator up the road, not hear about clean technology which might solve the problem in years to come.

Friends of the Earth, possibly because they do not wield as much money and power as Greenpeace, have not had the same impact but are regarded as a more community orientated organization. FoE have not made up their minds about toxic waste disposal. Some FoE members are unsure about incineration. Lack of funds appears to have stunted the growth of their excellent campaign against Britain's poisonous dumps, after a very exciting and credible beginning.

On one of the few occasions when Greenpeace and FoE achieved any real success in the opposition to toxic waste disposal, the impetus came from the community itself. Greenpeace may believe that ReChem would not have pulled out of Bonnybridge if it had not been for their involvement, but it is not the case. The people of Bonnybridge and Denny are grateful, but these people are more grateful to the tenacious local campaigners and to the workers who refused to handle PCB contaminated waste. "I think the reason why the plant closed, quite honestly, was because of the workforce. I've always said that it was the workforce who closed the plant and put themselves out of a job," said SCOTTIE's John Wheeler.[18]

The Green Party has not fared much better. Des Scholes, a former member of the Scottish Green Party, said there were too many "politicians" in the party. "What needs to be done," he said "is for the green movement to keep telling the truth in simple, straight-forward language".[9] Those in power at central level have been reluctant to relinquish their control, while the work at local level had not been capitalized on. Jane McNulty, a Green Party campaigner in Cadishead,

Greater Manchester, said after the 1990 local elections that the public regard the Greens as a one issue party. Yet Ray Jackson, who had campaigned against Lanstar and the importation of toxic waste to Manchester, won 8 per cent of the votes in Cadishead, beating the Liberals and getting double the Green vote of any other ward in Salford. "All his hard work on the Lanstar campaign may have paid off there and he will do better next time, I'm sure," McNulty noted.

There are indications that the green movement in Britain is developing in a similar way to the movement in the US. In the mid-eighties the green movement in the US had fragmented, largely because local green groups believed the national organizations were not interested in their "direct regional concerns".[20] Michael McCloskey warned of the "new splits developing in the environmental movement"; new "more militant" forces, he claimed, were emerging.[21] Kirkpatrick Sale noted that the environmental movement of the sixties in the US had been motivated by the anti-establishment mood of the decade, but then came a change: first there was an outcry for national regulations and palliatives; then a set of modest reformist laws; and finally, a whole new industry of environmental professionals – lobbyists, lawyers, publicists, bureaucrats and scientists. "Thus was born the new form of conservationism known as 'environmentalism' giving rise to the environmental establishment that has become so prominent in the last decade."[22]

Sale wrote that the charges against the "environmental establishment tend to cluster around four themes":

1. Environmentalists are reformist, working within "the system" in ways that ultimately reinforce it instead of seeking the thoroughgoing social and political changes that are necesssary to halt massive assaults on the natural world.
2. Environmentalists are anthropocentric, believing that the proper human purpose is to control and consume the resources of nature as wisely and safely – but as fully – as possible. They have yet to learn the ecocentric truth that nature and all its species have an intrinsic worth apart from any human designs.
3. Environmentalists have become co-opted into the world of . . . politics, playing the bureaucratic game like any other lobby, turning their backs on the grass roots support and idealism that gave the movement its initial momentum.
4. Environmentalists, finally, are not successful even on their own terms in protecting the wilderness, in stopping the onrush of industrial devastation. They are so caught up in compromise that they're actually going backward.[23]

What is really disturbing about these arguments is that they apply more directly to the British than to the US system. Sale argued that groups like Greenpeace could not be accused of bureaucratization and were unlikely to be seduced by the state. In Britain Greenpeace is very definitely bureaucratic and was seduced by the establishment fairly quickly. From a small grouping at the beginning of the eighties, Greenpeace today displays all the trappings of a multinational company or a civil service department. In his book *Green Warriors*, Fred Pearce argued that "Greenpeace must not hide behind its corporate persona and its market research. It ought to look afresh at what it thinks about the world".[24]

It has been left to individuals and communities to face the problems of toxic waste. Where groups have been able to avoid political differences (which irritates and drives away those with no party drum to beat) individuals have emerged to take on the various tasks and, more often than not, the campaigns have been relatively successful.

David Powell made this point about the campaign against ReChem in Pontypool.

> When the local group were involved the public must have had the view that somebody was looking after it for them. The group were almost all counsellors, which in my view was wrong. I became involved and started saying to people, nobody is looking after this and you better be aware of that, and it might encourage you to do something. There are so many issues going on all over the place that you can't expect people to be involved in all of them. But when there is a call for members of the public to show their support, you need a response which shows that the undercurrent of feeling is there. There have been some tremendous demonstrations and some phenomenal public meetings.[25]

It has been a frequent observation amongst anti-toxic campaigners that groups lose their impact when the campaign is hijacked by an individual or by an organization.

The discovery of the dioxin contaminated milk in Derbyshire and the publication of the British Medical Association's report should have prompted serious debate about toxic waste and human health, yet nothing has happened. Women's Environmental Network scientist Ann Link has suggested in her report on the dioxin-like compounds and human health that the pre-birth child is at serious risk from dioxin exposure. "It is likely that effects are already occurring at the high end of our exposure in the UK. We should be looking for these neurological effects, cutting down on sources, enabling women to decrease their body levels," she wrote.[26] Despite these events and the obvious concern among doctors and scientist, the apathy of both British and Irish governments on the toxic waste issue is remarkable. The 20th century has produced a legacy which is poisoning the planet, but we seem powerless to prevent it.

Notes and references

1. "New men at top as ReChem pair resign their jobs", *South Wales Argus*, 26 June 1991.
2. "BMA appeals on toxic waste", *Financial Times*; "BMA urges ban on toxic waste imports", *Guardian*; "Report calls for a waste plan after public health fear", *Scotsman*, 26 June 1991 See also British Medical Association, *Hazardous Waste and Public Health*, OUP, 1991.
3. Lamont, Tara; interview with Robert Allen, 1991.
4. House of Commons. Official report Col. 500 26 June 1991. See also MAFF press release FSD43/91, 26 June 1991.
5. Ibid.
6. Ibid.
7. Ibid.
8. Clark, Dvid; press release, 26 June 1991.
9. In May 1991 *Chemistry in Britain* ran a company profile of Coalite Chemicals, who employ 400 people at its only site, in Bolsover, Derbyshire. *Chemistry in Britain* said that the company had a turnover of £30 million per annum from a business divided into four areas: biocides, which contribute about 15 per cent of the profits; hydrocarbons and cresylics, which contribute 30 per cent; intermediates, 40 per cent and speciality chemicals, 15 per cent. "Coalite Chemicals uses coal-oil feedstock from the smokesless fuel site nearby and makes products as diverse as creosote (a low price, bulk product) and an intermediate used in the hair dye industry (an expensive, low volume product). The company also buys in raw materials from outside which it uses in other synthetic processes." *Chemistry in Britain* also noted: "Coal chemicals is traditionally a dirty and smelly business, but Coalite is committed to improving effluent and air emissions to minimise the effects of the company's processes on the environment." Coalite Chemicals, the author added, also had its own incinerator on site, which took only liquid waste. Further investigation revealed that Coalite were licensed to burn 120 gallons an hour of liquid chlorinated residues or approximately 5,000 tones a year. On 7 November 1991 the Government ordered Coalite to shut their incinerator down until modifications had been made.

10. Link, Ann; interview with Robert Allen, 1991.
11. British Medical Association, *Hazardous Waste and Human Health* (Oxford: OUP, 1991) pp. 138.
12. Lamont, Tara; interview with Robert Allen, 1991.
13. *Observer*, 7 July 1991.
14. Tinker, Jon; "Britain's Environment: Nanny knows best", *New Scientist*, 53, 789, p. 530
15. McCormick, John: *British Politics and the Environment* (London: Earthscan, 1991) p. 12.
16. Merck, Sharp and Dohme, the US multinational with a base in south Tipperary, were not actually prosecuted under existing law, they were taken to court by farmer John Hanrahan who won his case in the Supreme Court in 1988. Only Penn Chemicals and Angus Fine Chemicals have been prosecuted under the legislation. See also Allen and Jones, *Guests of the Nation* (London: Earthscan, 1990).
17. McCormick, op. cit.
18. Wheeler, J. Interview with Fiona Sinclair, 1991.
19. Scholes, D. Letter to the *Inverness Courier*, 7 June 1991.
20. Sale, Kirkpatrick. "The Forest for the Tress: Can today's environmentalists tell the difference?" *Mother Jones*, November 1986 pp. 25–33 and 58.
21. Ibid.
22. Ibid.
23. Ibid.
24. Pearce, Fred; *Green Warriors*, Bodley Head, 1991, p. 314.
25. Powell, D. Interview with Robert Allen, 1991.
26. *Women's Environmental Network, Chlorine, Pollution and the Parents at Tomorrow*. Contact WEN, Aberdeen Studios, 22 Highbury Grove, London N5 2EA, for further information.

Index

ABOUT WWF

WWF (World Wide Fund for Nature) is the largest private international nature conservation organization in the world supporting over 5,000 conservation projects in over 130 countries, 200 of them in the UK alone. WWF-UK is part of network of 23 national organizations working to protect our threatened environment. Of the funds raised in the UK, one third is used to fund WWF projects in the UK; the remainder, which is sent to WWF International in Switzerland, helps fund campaigns for tropical forests, marine conservation, protection of endangered species and habitats, Antarctica, bio-diversity and global warming.

For further information on the work of WWF write to: WWF-UK, Panda House, Weyside Park, Godalming, Surrey GU7 1XR.